THE SPURGE FAMILY
Euphorbiaceae

Cv
Cassava

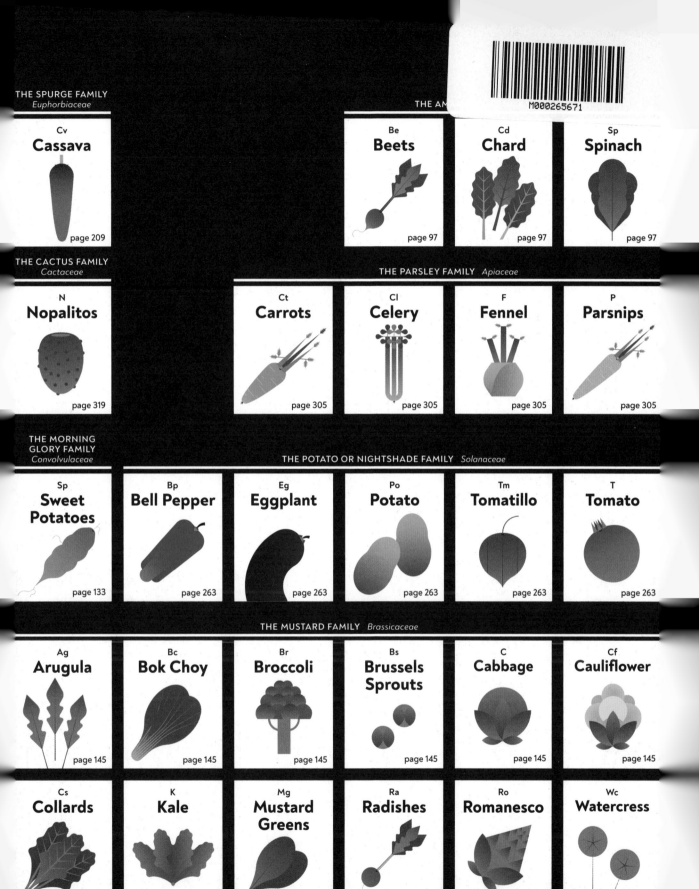

page 209

THE AM...

Be
Beets

page 97

Cd
Chard

page 97

Sp
Spinach

page 97

THE CACTUS FAMILY
Cactaceae

N
Nopalitos

page 319

THE PARSLEY FAMILY *Apiaceae*

Ct
Carrots

page 305

Cl
Celery

page 305

F
Fennel

page 305

P
Parsnips

page 305

THE MORNING GLORY FAMILY
Convolvulaceae

Sp
Sweet Potatoes

page 133

THE POTATO OR NIGHTSHADE FAMILY *Solanaceae*

Bp
Bell Pepper

page 263

Eg
Eggplant

page 263

Po
Potato

page 263

Tm
Tomatillo

page 263

T
Tomato

page 263

THE MUSTARD FAMILY *Brassicaceae*

Ag
Arugula

page 145

Bc
Bok Choy

page 145

Br
Broccoli

page 145

Bs
Brussels Sprouts

page 145

C
Cabbage

page 145

Cf
Cauliflower

page 145

Cs
Collards

page 145

K
Kale

page 145

Mg
Mustard Greens

page 145

Ra
Radishes

page 145

Ro
Romanesco

page 145

Wc
Watercress

page 145

Here are the big-on-flavor, vegetable-focused recipes acclaimed cookbook author, blogger, and food journalist Nik Sharma cooks nightly in his own kitchen. Sharma has put each of them "under the microscope," resulting in flavors and techniques that are tried and true. This cutting-edge cookbook, *Veg-table*, combines the science of *The Flavor Equation* with the warm, personal approach of *Season*.

Dive deep into vegetable families, from roots to legumes to the leafiest greens and beyond. Sharma reveals the origins, biology, and unique characteristics of 15 families of vegetables to help home cooks best choose, store, and cook them beautifully. Featuring more than 100 inventive and approachable recipes using vegetables both common and uncommon, *Veg-table* inspires flavorful and nutritious home cooking for anyone looking to diversify their dinner plate.

Dishes range from simple weeknight pastas—Pasta with Broccoli Miso Sauce, Shallots + Spicy Mushroom Pasta—to vegetable-focused meals that aren't strictly vegetarian: Flex your healthy diet with Swordfish + Crispy Cassava with Chimichurri or Chicken Katsu with Poppy Seed Coleslaw. A wide variety of hot and cold soups, salads, sides, sauces, and rice-, egg-, and bean-based dishes round out this collection.

Sharma is a molecular biologist, beloved food blogger, contributor to *Serious Eats*, and the author of the buzz-generating, James Beard– and IACP-nominated cookbooks *Season* and *The Flavor Equation*. Sharma is famous for his deeply flavorful and unique recipes, inspired by his Indian heritage and time in the American South, and his deep love of cooking.

Veg-table's organization, research, and thoughtful curation signal a new kind of reference cookbook. More than 100 of Sharma's gorgeous and evocative photographs accompany a dozen instructive illustrations to help those who learn best visually.

Veg-table is an exciting new reference from a major talent that will round out your cookbook collection, one that you'll turn to again and again, in every season, year after year.

VEG-
TA
BLE

VEG-TABLE

**Recipes, Techniques +
Plant Science for Big-Flavored,
Vegetable-Focused Meals**

Recipes and Photographs by
NIK SHARMA

CHRONICLE BOOKS
SAN FRANCISCO

For Gus, the world is your vegetable.

For Michael, Paddington, Vesper, and Drogy with love.

Library of Congress Cataloging-in-Publication Data

Names: Sharma, Nik, author.
Title: Veg-table : recipes, techniques + plant science for big-flavored,
   vegetable-focused meals / recipes and photographs by Nik Sharma.
Description: San Francisco : Chronicle Books, [2023] | Includes index.
Identifiers: LCCN 2023020292 | ISBN 9781797216317 (hardcover)
Subjects: LCSH: Cooking (Vegetables) | LCGFT: Cookbooks.
Classification: LCC TX801 .S46 2023 | DDC 641.6/5--dc23/eng/20230502
LC record available at https://lccn.loc.gov/2023020292

Manufactured in China.

MIX
Paper from responsible sources
FSC™ C169962
FSC
www.fsc.org

Food and prop styling by Nik Sharma.
Design by Lizzie Vaughan.
Typesetting by Frank Brayton.
Illustrations by Matteo Riva.

All-Clad is a registered trademark of All-Clad Metal Crafters, LLC; Aroy-d is a registered trademark of Thai Agri Foods Public Company Ltd.; Beano is a registered trademark of Medtech Products Inc.; Gas-X is a registered trademark of Gsk Consumer Healthcare LLC; Green Valley Creamery is a registered trademark of Redwood Hill Farm & Creamery, Inc.; Instant Pot is a registered trademark of Instant Brands Inc.; Kevlar is a registered trademark of Dupont Safety & Construction, Inc.; Kikkoman is a registered trademark of Kikkoman Corporation; Kong Yen is a registered trademark of Kong Yen Corporation; Lifeway is a registered trademark of Lifeway Foods, Inc.; Mekhala is a registered trademark of Mekhala Pte. Ltd.; Microplane is a registered trademark of Grace Manufacturing, Inc.; New York Shuk is a registered trademark of New York Shuk LLC; Oaktown Spice Shop is a registered trademark of Oaktown Spice Shop, LLC; OXO is a registered trademark of Helen of Troy Ltd; *Star Wars* is a registered trademark of Lucasfilm Ltd.; Zwilling is a registered trademark of Zwilling J.A. Henckels.

10 9 8 7 6 5 4 3 2 1

Chronicle books and gifts are available at special quantity discounts to corporations, professional associations, literacy programs, and other organizations. For details and discount information, please contact our premiums department at corporatesales@chroniclebooks.com or at 1-800-759-0190.

Chronicle Books LLC
680 Second Street
San Francisco, California 94107
www.chroniclebooks.com

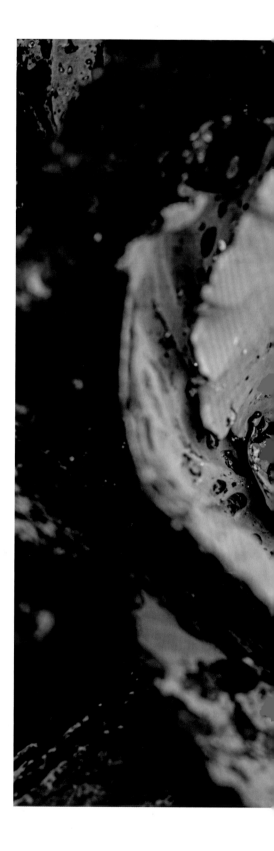

Table of Contents

1.
Onions, Shallots, Scallions, Leeks, Garlic + Chives
page 31

2.
Yams
page 53

3.
Bamboo + Corn
page 63

4.
Asparagus
page 83

5.
Beets, Chard + Spinach
page 97

6.
Artichokes, Sunchokes, Endive, Escarole, Radicchio + Lettuce
page 113

Introduction

"Can we get that little chilli plant? Please, please, please?"

WORLD MAP OF CULTIVATED PLANT ORIGINS

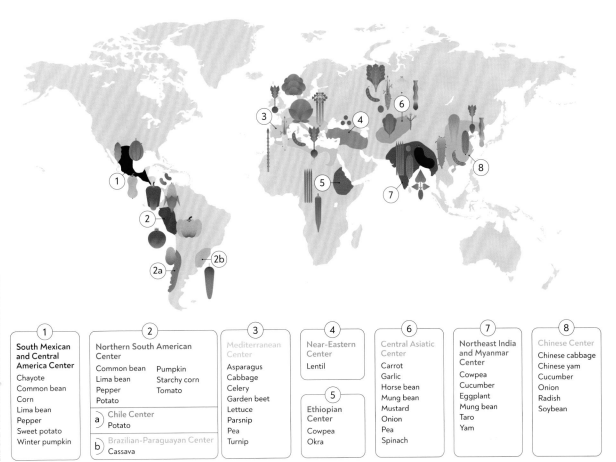

Adapted from Gideon Ladizinsky's *Plant Evolution under Domestication* and G. E. Welbaum's *Vegetable Production and Practices.*

① South Mexican and Central America Center
Chayote
Common bean
Corn
Lima bean
Pepper
Sweet potato
Winter pumpkin

② Northern South American Center
Common bean
Lima bean
Pepper
Potato
Pumpkin
Starchy corn
Tomato

ⓐ Chile Center
Potato

ⓑ Brazilian-Paraguayan Center
Cassava

③ Mediterranean Center
Asparagus
Cabbage
Celery
Garden beet
Lettuce
Parsnip
Pea
Turnip

④ Near-Eastern Center
Lentil

⑤ Ethiopian Center
Cowpea
Okra

⑥ Central Asiatic Center
Carrot
Garlic
Horse bean
Mung bean
Mustard
Onion
Pea
Spinach

⑦ Northeast India and Myanmar Center
Cowpea
Cucumber
Eggplant
Mung bean
Taro
Yam

⑧ Chinese Center
Chinese cabbage
Chinese yam
Cucumber
Onion
Radish
Soybean

I begged and harassed my parents incessantly, knowing all too well that growing this plant from the nursery would prove to be a big challenge for me. I didn't even like hot chillies at that point in my life, so my enthusiastic desire was a bit bizarre. The romantic notion of growing my peppers on our tiny windowsill in Bombay, India (even though the name of the city changed to Mumbai, it will forever remain Bombay to me), had fogged my mind with a cloud of desire. But experience should have taught me otherwise. I had already tried to grow wheat, chickpeas, and rice for my geography class and Christmas manger scenes. The plants grew at first but never lasted. Still, for a child barely into his teens, to see a dried seed transform and sprout into something alive and growing was fascinating.

The chilli plant came home. It lived for a week. It died.

Years passed. Plants came and went. My interests developed and, not surprisingly, my love of seeing things grow and live meant I fell in love with biology in school. I majored in microbiology and biochemistry and moved to the United States for graduate school to study molecular genetics and later public health policy.

No matter what academic program I was in, I learned that food played an important role in both the prevention and development of disease. The components inside foods and ingredients were also the basis for many experiments that we used to study diseases. For example, we used the

11

lectin proteins from legumes to study and separate special sugar-tagged proteins called *glycoproteins* from blood. Turmeric was used to detect increases in pH in lab titrations when alkalis such as sodium carbonate were added to an acid like acetic acid. We used beads made from either silica or agarose (a type of polysaccharide obtained from red algae) to bind enzymes, alcohols, and fats to extract the essential oils from fruits and vegetables, and we used sugar and salt to help transport molecules across membranes. It was this underlying thread of biological molecules and their actions in the lab combined with their presence in my everyday life that drew me to cooking and led to my transition from science to the kitchen and becoming a cook and recipe writer.

Then, as now, I lived far away from home, oceans and continents away from India. With my newfound freedom, I started to indulge my romantic notions of gardening. I moved into an apartment and, with space all to myself, grew my own plants. But my confidence was weak. My personal history was of two- to three-week plant-growing successes. For a house-warming gift, I received one of those extremely popular "lucky bamboo" plants from a friend. It needed nothing but water. I kept it in the kitchen next to the sink, as an easy reminder. Over time, as the roots grew longer and the plant taller, I eventually moved it to a bigger container to water it. And it lived!

With this success, I grew more confident. I had kept a plant alive! As I got braver, then came the cacti and succulents. I tried my hand at easy-growing mint and pepper plants. They'd grow and the satisfaction of growing something that I could eat brought me immense joy. The child who couldn't grow a chilli plant in Bombay now grew several varieties and learned to use and appreciate their flavors.

Now I spend as much time gardening as I do cooking. Southern California is warm for most of the year, with a short rainy season. When I lived in Oakland, California, I kicked off my first proper, large-scale excursion into edible gardening. Our backyard was small but offered just enough space for a novice like me. My husband, Michael, and I dug out a foot of the old, worn-out soil and replaced it with fresh, rich compost. I sought help from Leslie Bennett, a designer of edible landscapes, who helped me select plants that worked with the light in our yard. I grew peppers, tomatoes, tomatillos, various kinds of citrus, and even passion fruit. I also learned to loathe squirrels, who were way too intelligent for me and adept at stealing my ripe figs.

Life always brings change, and we moved to Los Angeles, where the weather is much warmer and drier. Our new home has a larger backyard, but when we arrived it was in messy shape. The landscape was covered with thickly intertwined brambles of lantana, which I ripped out. On this blank canvas, I designed the backyard for growing everything I liked to eat and cook with.

Dwarf fig, lemon, lime, and mandarin orange trees grow where the lantana once rambled. Under the thick heat of summer, scallions, onions, peppers, and tomatoes thrive. Growing my own produce is a privilege, one that I don't take lightly. My garden gives the space and opportunity to grow ingredients from India, such as the drumstick tree (also known as moringa), and also experiment with new-to-me produce, such as cactus pads and finger limes. I've got a few curry leaf plants that remind me of India, housewarming gifts from Hugh Merwin, my friend and fellow gardener and writer. Any recipes that contain curry leaves you come across here I developed with leaves from my own garden.

Like the kitchen, the garden has become my lab, and the ingredients I grow make their way into the decisions I make in my daily cooking and the recipes I develop for work. I tinker with new plant varieties and soil conditions, attempt cross-pollination between plants to create new varieties, and pursue more experimentation. It's chaotic yet fun.

The vegetables and fruits I harvest are so much tastier than ones from the market. Using them teaches me to adjust my seasonings and my culinary techniques. For example, my homegrown bell peppers are juicier and sweeter than the ones I pick up from the store, so I compensate by using less salt when I cook them. I've also learned that some ingredients are best obtained from professional growers: For example, my twenty chickpea plants barely yielded 1 cup [160 g]. Gardening failures make me appreciate our farmers even more.

This is a book about vegetables, some more familiar than others, but all equally satisfying. Dive in, and learn more about those you're familiar with and those that feel new to you. Use them in fun and exciting ways in your kitchen. My intention in this book is to give you techniques, flavors, and ideas—with foundations in science and in history—to become an inventive and frequent vegetable home cook. Now, let's start cooking!

What's a Vegetable?

I admit that I still use a terribly faulty definition of a vegetable when I'm cooking with plant ingredients: If it tastes sweet, it's a fruit; otherwise, it's a vegetable. The definition of a vegetable changes depending on who is defining it; the concept of a vegetable is fluid. From a botanist's viewpoint, the definition is very precise: Fruits are the mature ripened ovaries of the plant that forms from the plant's flowers, and this includes many foods we call vegetables, like okra, tomatoes, eggplants, cucumbers, and so on. Even nuts that are enclosed by shells, like coconuts, pistachios, and walnuts; grains like rice and barley; and spices like vanilla beans and black peppercorns are considered fruits.

However, when it comes to vegetables, things get a bit subjective. A vegetable is usually the fruit of a plant (leaves and flowers are exceptions). The word *vegetable* itself is not a botanical term; it is generally applied according to how the plant is used. Another way to think about what makes a fruit a fruit and a vegetable a vegetable is this: Fruits develop from the flower of a plant (think apple or tomato), while vegetables often also include different edible parts of the plant, such as tubers (potatoes and sweet potatoes), leaves (spinach and chard), bulbs (fennel), stems (celery), or budding forms (cauliflower, broccoli). But a cook sees a vegetable when eggplant meets the searing heat of a wok (never mind that its strawberry hull–like stem cap and internal seeds are clearly hallmarks of fruit) or in the handful of cremini mushrooms incorporated into an omelet (never mind that mushrooms aren't really plants, but fruiting bodies from the fungi kingdom).

But in terms of cooking, and an overall approach to vegetable cooking in your kitchen, what makes one vegetable like or unlike another? The answer to this question lies in how we approach it: Vegetables can be sorted by a variety of factors, from their origin to how we use them, how they like to grow (weather, soil conditions, water needs, etc.), their life span (annual, biennials, or perennial crops), their edible parts, physical features, and even their genes.

How does all of this inform our cooking— and why are definitions even relevant?

When I'm shopping for produce at the market or grocery store, developing a recipe, or planning a meal at home, the different categories of vegetables play a major role. Root vegetables, such as sweet potatoes and parsnips, will need a bit more time to cook until soft and tender. Some vegetables—like yams, bamboo shoots, and cassava—must be cooked completely before eating because they contain naturally occurring toxic chemicals that heat destroys.

Produce categories also factor into deciding how much I should buy—what I can expect to use, and whether I should buy extra to use another day. Seasonality dictates what I can cook and when; for example, it's fresh ripe tomatoes in summer and canned tomatoes or tomato paste in the cooler months of the year when I need their flavor. And if I buy fresh spinach, I need to buy plenty, because it shrinks down dramatically. A vegetable like an eggplant needs to be used soon because it spoils easily, whereas potatoes and onions can keep well, but they need a cool,

dry, and dark place to make them last longer.

Vegetables come in such a wide variety of colors and textures. Adding unexpected varieties of familiar vegetables—likes purple bell peppers and zebra tomatoes in a fresh salad—brings visual excitement to a meal. An assortment of textures in a dish also prevents palette fatigue—the diner won't grow weary of tasting the same flavors, smelling the same aromas, and eating the same monotonous textures repeatedly in one sitting. Adding crunchy toasted nuts and a sprinkling of microgreens to tender roasted sweet potatoes, or incorporating chunks of squash or beans in a soup, for example, brings interest to a meal and excites the senses. Of course, it can also provide an interesting source of conversation with dinner companions: "Why did you choose okra for these tacos (page 255) and pair them with the butter chicken sauce?" or "This is my first time tasting this vegetable! I didn't expect that taste or texture, and I love it!"

But when you're at the produce section, staring at a diverse display and wondering what to buy, why do any of these factors matter? Whether you're combining vegetables into one dish or soloing with one star element, knowing what to use and how to use it opens up a whole world of different flavors, textures, and choices. Vegetables can be categorized in a multitude of ways, from their origin to the parts we eat and how we eat them, their growing season, and more. Let's dive into some of these categories.

Birthplace: Did this vegetable originate in the Western or Eastern Hemisphere?

The arrival of the Europeans in the Americas led to the introduction of new plants and vegetables to both Afro-Eurasia (Europe, Asia, and the Middle East) and the Americas (North and South America). Peppers, potatoes, pumpkins, and tomatoes are just some of the many vegetables that originated in the Americas and quickly went on to become a part of many cuisines across the globe. It's always intriguing both academically and practically to see how ingredients like peppers, potatoes, and tomatoes became an integral part of the identity of many different cuisines around the world. A single ingredient gets transformed and used in different ways depending on how a culture looks at it.

Consider tomatoes, used to make the base of sauces and stews in Italy, India, and many other parts of the world. But in Japan, tomatoes are used in garnishes, in grilled yakitori preparations, and in yoshoku preparations. Yoshoku refers to Western food that's been reinterpreted to suit Japanese tastes, such as napolitan or naporitan, a spaghetti dish made with ketchup or tomato sauce, bell peppers, sausage, garlic, and mushrooms.

Are they fresh or processed?

One way to categorize vegetables is based on the way they get to our tables. Fresh vegetables are minimally processed; they go straight from the ground to the market or grocery stores and then to our kitchens (if you grow them in your own garden, then it's one stop fewer). Precut vegetables, like the diced onions and carrots sold at the grocery store and shelled fresh beans and peas, also fall into this category because they aren't cooked, and the processing is minimal. On the other hand, vegetables such as peas, spinach, beans that are sold canned, and those that are dried, pickled, frozen, or made into pastes or powders are considered processed. These vegetables go through several different stages of treatment to ensure their longevity. Does processing make them less palatable or nutritious? It depends on a number of factors.

Convenience is a big reason why these methods exist. Efficiency is another: These technologies enable us to feed more people with the same amount of food. In some cases, processing a vegetable might improve storage and preserve the nutrient levels more reliably than if they were kept fresh. Certain vegetables, like cucumbers, tomatoes, and asparagus, aren't known for their longevity. Nutrients like vitamin C (ascorbic acid) and carotene (the source for vitamin A) decrease dramatically with time.

Once a vegetable ripens and is picked, enzymes within the plant cells go into full gear and begin to tear down the starch and protein, destroying the vegetable's quality. Vitamins B and C decrease with cooking through exposure to heat; in fact, the changes in vitamin C levels are often used to measure the quality of foods, whether fresh, in storage, or during cooking. Some nutrients, like the fat-soluble vitamins A, D, E, and K, present in vegetables such as carrots and sweet potatoes, are better eaten combined with a fat, like the olive oil in a vinaigrette or cooking; the fat aids more efficient absorption of these vitamins into our body. There are some vegetables, like cassava and yams, that must be processed by cooking to destroy the naturally present toxic chemicals and render them safe to eat. So, processing food isn't always bad; as with all things, the context matters.

FRESH VEGETABLES

 Short shelf life

 Ready to eat

 Needs little to no prep work before eating

PROCESSED VEGETABLES

 Longer shelf life

 Might need to be prepped before use

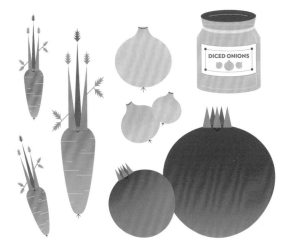

DICED ONIONS

PICKLES

FROZEN SPINACH

DRIED BEANS

TOMATO PASTE

CANNED BAMBOO

The growing seasons

It's true that vegetables taste best when picked at the right time, when they're in the prime of their growing season. The seasons not only provide the plant the necessary and appropriate conditions to grow efficiently but also help them ripen properly to taste their best. Not surprisingly, this might just be one of the oldest and most common ways to sort vegetables. Buying vegetables when they are in season will make your dishes taste more flavorful, making the most of the spices and other accoutrements that you add to enhance the vegetables' performance. For out-of-season times when you nevertheless want to use those vegetables, processed and preserved options like canned, pastes, and the like will get the job done (subject to your recipe).

Many farmers and gardeners use companion planting, a practice in which certain combinations are grown together. The plants provide nutrients to one another, offer mutual protection from pests and harsh weather, and attract pollinators. Many vegetables that are grown as companions also come together wonderfully in the kitchen. Tomatoes and garlic are planted together because the garlic protects the tomato plant from mites and aphids. We see this combination extend to the kitchen, where a velvety marinara sauce made from ripe tomatoes comes to life with the warm bite of garlic (try the Spaghetti with Roasted Tomato Miso Sauce, page 301).

VEGETABLES BY THEIR PREFERRED GROWING SEASON

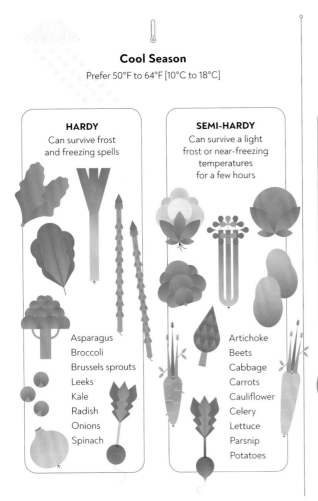

Cool Season
Prefer 50°F to 64°F [10°C to 18°C]

HARDY
Can survive frost and freezing spells

Asparagus
Broccoli
Brussels sprouts
Leeks
Kale
Radish
Onions
Spinach

SEMI-HARDY
Can survive a light frost or near-freezing temperatures for a few hours

Artichoke
Beets
Cabbage
Carrots
Cauliflower
Celery
Lettuce
Parsnip
Potatoes

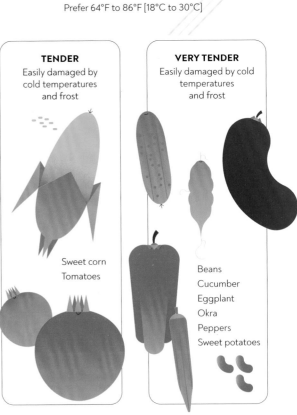

Warm Season
Prefer 64°F to 86°F [18°C to 30°C]

TENDER
Easily damaged by cold temperatures and frost

Sweet corn
Tomatoes

VERY TENDER
Easily damaged by cold temperatures and frost

Beans
Cucumber
Eggplant
Okra
Peppers
Sweet potatoes

Adapted from G. E. Welbaum's *Vegetable Production and Practices.*

Edible parts

There are some vegetables—sweet potatoes and beets are two—that offer edible tubers and greens; for others, like eggplant, only the fruit is edible and the rest of the plant is toxic. Vegetables can be defined and sorted by the parts of the plant we eat. Starchy vegetables that grow underground, such as potatoes and sunchokes, need to be cooked before eating; roots like beets and carrots can be eaten raw or cooked, and greens can be eaten raw or cooked.

Roots	Beets, carrots, radish, parsnips
Stems	*aboveground:* Asparagus, celery
	underground: Ginger, potatoes, onions, yam, taro
Tubers	*enlarged underground stem:* Potatoes, sunchokes
Corm	*base of the stem:* Taro
Leaves	Alliums, collards, lettuce, onions, radicchio, scallions, spinach
Flowers	Broccoli, cauliflower
Fruits	Cucumbers, squash, pumpkins, tomatoes, peppers

VEGETABLES BY THE PARTS WE EAT

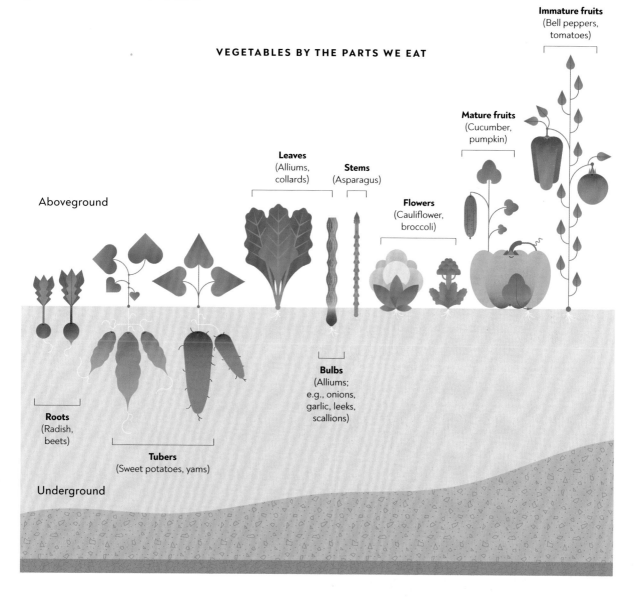

Aboveground

Immature fruits
(Bell peppers, tomatoes)

Mature fruits
(Cucumber, pumpkin)

Leaves
(Alliums, collards)

Stems
(Asparagus)

Flowers
(Cauliflower, broccoli)

Bulbs
(Alliums; e.g., onions, garlic, leeks, scallions)

Roots
(Radish, beets)

Tubers
(Sweet potatoes, yams)

Underground

17

Plant science families

The botanist Liberty Hyde Bailey grouped plants into four categories: algae and fungi, mosses and liverworts, ferns, and seed plants (which primarily include most vegetables we eat). Most people consider fungi, such as mushrooms, as vegetables. Algae—think nori, laver, and kombu—are sometimes called sea vegetables. Mosses such as reindeer moss and ferns such as fiddlehead ferns are also cooked and eaten. For culinary purposes, this plant classification matters because knowing the plant class will help with buying, storage, prepping, and cooking. Plants that belong to the same family will have similar growing and harvesting seasons, similar flavor profiles, and similar textures. Consequently, they can be prepared and cooked using the same methods. For example, Solanaceae (tomatoes and bell peppers) taste best in summer and are often paired together in many dishes, like Gazpacho (page 270). In this book, the recipes are focused on the seed plant family, but you will notice some members of the other plant families making an appearance (Leek + Mushroom Toast, page 42). For a more detailed overview of how vegetables can be classified by their plant families, check out the endpapers at the front and back of the book.

Monocotyledoneae

SEED CONTAINS A SINGLE LEAF

Amaryllidaceae
Amaryllis family
Chives, garlic, leeks, onion, scallions, shallots

Dioscoreaceae
Yam family
Ube, yams

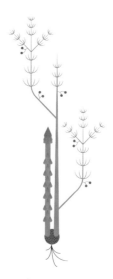

Asparagaceae
Asparagus family
Asparagus

Poaceae
Grass family
Bamboo, corn

Dicotyledoneae

SEED CONTAINS TWO LEAVES

Amaranthaceae
Amaranth family

Beet, chard, spinach

Asteraceae
Sunflower family

Artichoke, endive, escarole,
lettuce, radicchio, sunchoke

Convolvulaceae
Morning glory family

Sweet potatoes

Solanaceae
Potato or nightshade family

Bell peppers, eggplant,
potatoes, tomatillos, tomatoes

Malvaceae
Mallow or cotton family

Okra

Euphorbiaceae
Spurge family

Cassava

Fabaceae or Leguminosae
Pea or bean family

Beans, chickpeas, green beans,
jícama, lentils, peanuts, peas

Cactaceae
Cactus family

Nopalitos

Apiaceae
Parsley family

Carrots, celery, fennel, parsnip

Brassicaceae
Mustard family

Arugula, bok choy, broccoli, Brussels
sprouts, cabbage, cauliflower, collards, kale,
mustard, radish, Romanesco, watercress

Cucurbitaceae
Gourd family

Chayote, cucumber,
pumpkin, squash

The Veg-Table

VEGETABLE	SMALL	MEDIUM	LARGE	XL
ARTICHOKES		130 g	455 g	
AVOCADO	150g	185 g	220 g	
BEETS	70 g	120 g		
BELL PEPPER	100 g	130 g	180 g	200 g
BUTTERNUT SQUASH			1.3 kg	
CABBAGE				
GREEN		910g	1.4 kg	
SAVOY		700g		
CARROTS	50 g	60g	70 g	
CAULIFLOWER	250 g	400 g	800 g	
CELERY (STALK)	18 g	30 g	40 g	
CHAYOTE		230 g		
CUCUMBER				
ENGLISH	95 g	120 g	340 g	
PERSIAN		85 g		
GARLIC (HEAD)	30 g	50 g		
EGGPLANT				
BABY	100 g			
GLOBE		370 g	550 g	
JAPANESE		230 g		
ENDIVE		85 g		
FENNEL (BULB)	170 g	200 g	270 g	
JÍCAMA			455 g	
LEEKS	25 g	150 g	270 g	
LETTUCE				
BABY GEM (HEAD)		150 g		
ROMAINE (HEAD)		170 g		
ONIONS	150 g	200 g	300 g	400 g
PARSNIPS	40 g	45 g	55 g	
POTATO				
RUSSET		300 g	400 g	
YUKON GOLD		300 g	400 g	
PUMPKIN, SUGAR PIE	910 g			
RADICCHIO	85 g	100 g	340 g	
SHALLOTS				
SINGLE BULB	35 g	40 g	45 g	
DOUBLE BULB	55 g	65 g	80 g	
SWEET POTATOES	150 g	250 g	400 g	
YELLOW SQUASH		200 g		
ZUCCHINI		85 g		

NOTE: These weights are based on average vegetable weights I calculated from produce obtained from different grocery stores in the United States. The USDA uses dimensions rather than weight to grade vegetables.

In the Kitchen

Cooking should be a pleasant experience, not an intimidating one. Somehow, we've all subscribed to the notion that every dish that's home-made or from-scratch is superior. This is not always true. Do what is comfortable for you and makes your work less stressful in the kitchen. If you don't want to make a spice blend from scratch, that's OK! I buy these, too. My friends at Oaktown Spice Shop can make a garam masala blend that tastes different—and exciting!— from what I would come up with. Look for a trustworthy source of spices and purchase a jar from them.

I keep boxes of store-bought (Hi, Ina Garten!) salted and unsalted toasted nuts and seeds such as cashews, almonds, pumpkin seeds, and so on in my freezer for my cooking needs. It's a waste of time and energy to pull out a skillet or turn on the oven to toast a small amount, say 2 Tbsp, of nuts or seeds. I toast a large batch and freeze it; you could also purchase toasted nuts.

If you need to use canned, frozen, or prechopped vegetables from the grocery store, that's fine too. There are times when we're either too tired or simply don't have the time to fol-low every fresh, DIY step in a recipe. Take shortcuts when it suits you to make life in the kitchen just that much easier—you may end up enjoying the time you spend there more fully. This cookbook is not meant to be an exercise in superior methods and ingredient sourcing. I'm a home cook, like you.

Perhaps it's LA working on me, but I like to think of recipes as a movie set. Think of how the actors (in this case, the key ingredients) perform through a story line. For ingredients, rather than plots and relationships, you have cooking techniques and flavor boosters affecting the tra-jectory. I define "flavor boosters" as any ingredient that can influence what you taste in a dish. Think sea salt, the spice mixes in your pantry, the bottle of caramel-colored black vinegar from the Chinese grocery store. Flavor boosters can also be techniques that add flavor, such as charring an eggplant or zucchini on the hot grates of a grill to get that smoky essence in the dish. They're all there to help you transform your ingredients.

A Few General Cooking Tips

- I've provided a chart to be used with the recipes in this book, which includes the average weights (that I've determined) for the most common sizes of vegetables found in grocery stores, to help you get an estimate of what you need in the recipes. If you have considerably more or less of the main ingredient, taste and season as you go.

- Water is an enemy of roasting. Wash your vegetables and dry them well before sticking them into the oven. Toss them well with oil or fat with a high smoke point. (The smoke point of the oil should be greater than the oven temperature used. For most applications, like roasting and frying, extra-virgin olive oil (325°F to 375°F [165°C to 190°C]) and grapeseed oil (smoke point 485°F [250°C]) work great. I prefer grapeseed oil for frying because of its neutral flavor, and keep in mind that extra-virgin olive oil might leave a slight aftertaste that won't work with all recipes.

- If using spices on your roasting vegetables, watch them carefully because they can char depending on the temperature and the cooking time.

- Spread out the vegetables on the pan if you want them to become crisp; proper air circulation is essential for evaporation of water, or the vegetables will end up sitting in a pool of their own liquid and become soggy rather than crisp. If the pan is overcrowded, divide the vegetables between two pans to cook; trying to cook too much in one pan will result in uneven cooking or sogginess. For larger pieces of vegetables like broccoli and cauliflower florets or sweet potatoes cut into wedges, place a lightly greased wire rack on the baking sheet and add the vege-tables on top of the rack, allowing air and heat to circulate all around the florets, resulting in maximum crispiness and even browning.

- For efficient and even cooking, always rotate the roasting dish or baking sheet halfway through the cooking period.

- Vegetables such as squash and eggplant should be salted and left to sit before roasting, to draw excess moisture from their flesh.

Before cooking, rinse off the salt and pat dry for a crispier texture.

- Vegetables that are rich in sugar, such as onions and sweet potatoes, can char at high temperatures, so watch them carefully when frying or roasting.

- When steaming vegetables, season them with spices *after* cooking, or else the steam will wash away the spices as the water condenses.

- When steaming vegetables, line the base of the basket with vegetable leaves such as cabbage or lettuce instead of paper. It's more flavorful, is environmentally friendly, and reduces waste.

- Pectin is a carbohydrate that makes vegetables firm to the touch by holding their cells tightly together. How pectin is manipulated during cooking affects the outcome. Baking soda in boiling water softens pectin; when cooking starchy vegetables like yams, cassava, and potatoes, baking soda will remove the pectin, releasing some of the cells and starch granules. This method is employed to produce a crunchy and crispy coating on roasted potatoes and cassava (see Cassava Bravas, page 212).

- Caramelization and the Maillard reaction are the two major flavor-building reactions that come into play during heat-based cooking methods. Bittersweet flavors, toasty and smoky aromas, and dark-brown caramel color are the hallmarks of these reactions. Caramelization occurs between sugars as the Maillard reaction occurs between sugars and the amino acids of proteins. Note that both reactions can occur simultaneously, but the degree to which each occurs is a more complex question, beyond the scope of this book. Water is essential for these reactions to occur, but too much water slows these reactions. Vegetables that are stir-fried, seared, roasted, or grilled will show a pronounced development of flavor and color compared to vegetables that are boiled or steamed. It's not that those reactions aren't taking place; they are, but at a snail's pace in comparison. Baking soda is a catalyst for these reactions, so adding a little bit will help accelerate these flavor-building reactions.

A Few of My Favorite Kitchen Tools

The frugal and clutter-averse side of me avoids buying and collecting kitchen gadgets with limited utility. However, there are some kitchen tools that I'm all for because they make my life easier and serve more than one purpose. The following list is by no means exhaustive, but rather highlights a few tools that I turn to often.

Knives: I use an 8 in [20 cm] chef's knife and a 4 in [10 cm] paring knife. These two are all you'll need for most of your vegetable prepping needs. A serrated knife is good for delicate vegetables like tomatoes, because the serrated edge produces a clean cut using very little pressure. Regardless of what knives you use, make sure they're sharp and be careful.

Vegetable peeler: Besides peeling off the skin of cucumbers and potatoes, I use my peeler to shave cabbage into thin shreds for coleslaw and salads. Get the Y-shaped peeler; its wider angle helps shave vegetables more efficiently.

Citrus zester and Microplane zester: I use both types of zesters depending on the need. As its name implies, a citrus zester works great to grab thin strips of zest from lemons, limes, and other citrus fruits for cooking or garnish. You can also use it to score vegetables and create various shapes and patterns. For example, I use my citrus zester to channel my carrots before I slice them on the mandoline so the slices resemble flowers, or to get those pretty striped edges on cucumbers and eggplants (and lose enough peel that the cuke slices aren't bitter but also hold together and show a bit of green). I reserve my Microplane zester, aka rasp grater, for when I need very fine zest and don't need the peel to show up visually or as a texture in what I'm making, such as in a stew or soup. The rasp grater is also great for producing fine, powdery shavings of cinnamon and nutmeg, or finely grating garlic, ginger, and even cheese.

Ricers and food mills: Besides making terrific mashed potatoes, use these for anything else you need to mash or purée, including broccoli, carrots, and tomatoes. I use my OXO ricer for mashed potatoes, and my All-Clad food mill (which I picked up at an outlet sale) is my go-to for large quantities of potatoes and tomatoes; it also gives a smooth texture. Food mills take up more storage space than ricers, and if you can't be bothered to keep either of these tools, you can

22

mash soft-cooked vegetables like potatoes with a fork and then press them through a fine mesh strainer to get a smooth texture.

Mandoline: I use this tool sparingly because I'm accident-prone. But nothing beats the perfect, thin, consistent slices of fennel, potato, carrot, cucumber, or apple you can achieve with a mandoline. Get a pair of Kevlar gloves to wear when working with it, because the blades are extremely sharp.

Steamer basket: I keep a bamboo steamer to steam dumplings and vegetables (a stainless steel one works just as well). They're easy to use and maintain and need only to be dried after use. Before steaming, line the base of the basket with cabbage or lettuce leaves or sheets of parchment paper to prevent food—especially dumplings—from sticking.

Salad spinner: I resisted adding a salad spinner to my kitchen for years, until I received one for Christmas. It's a drier that relies on centrifugal force (because it spins at a high speed, water and other liquids get tossed out) to get rid of excess water for the crispest salad greens and clean, dry herbs. You can also use the container to keep undressed washed and dried salad fresh in the refrigerator; it will function as an additional crisper storage in the fridge.

Pastry brush: Over the past couple of years, I've come to like silicone pastry brushes for applying a thin coating of any liquid to food, be it an egg wash over buns or oil over vegetables to be grilled or roasted (see Hasselback Parsnips with Pistachio Pesto, page 310). Unlike the brushes made from synthetic or animal hairs, silicone brushes never leave bristles behind in food.

Cast-iron spice grinder: Unlike an ordinary pepper mill, which needs to be replaced often, a cast-iron spice grinder will last for years. I keep one to grind pepper and another to grind spices.

Offset spatula: Not only are these useful for frosting cakes, but they're also wonderful for spreading even layers of spices or pastes onto vegetables (see Collards Patra, page 148). They're also great for getting into the tight corners of pans or picking up small, delicate items, such as cookies.

Food processor or high-speed blender: A food processor is great when you need to grind an ingredient to a coarse texture; it gives you better control and regulates the speed for a few short seconds, and the blades can be changed. A high-speed blender is wonderful for making emulsions, creamy soups, smoothies, and the smoothest dips and spreads.

How to Use This Book

For those who have cooked through my previous books or columns, you will notice a big departure in my style of recipe writing in this book. Based on the feedback I've received directly from my readers over the years, I've streamlined my recipes to make them easier to read and follow. Taking a cue from the legendary *The Joy of Cooking*, I've listed ingredients within the recipe steps. The ingredients and their amounts are bolded within the steps so you can easily see them while cooking, with the book propped on your counter. This is exactly how I write recipes when I develop them in my kitchen. I've also tried to minimize the use of extra cookware so cleaning up is less cumbersome.

I've included The Vegetable Pantry (page 26), which I hope you will find helpful. I encourage you to visit the Postharvest Center at the University of California, San Diego for their excellent resources and information on storing produce.

Before you embark on a recipe, I urge you to first read the Cook's Notes. There you will find tips, explanations of why and how things work, and information on tweaking the techniques, making substitutes, and other ways of exerting your personality in these recipes.

A Few Words on Substitutions: This isn't a book on desserts, and there's no baking (just roasting), making it

very easy to substitute ingredients based on your needs. However, when you make a substitution, always pay attention to your cooking times and seasonings, and adjust them accordingly. In the recipes for some of the vegetables, such as Yams (see page 53), you can swap in regular potatoes, but adjust the cooking conditions as needed.

This is by no means a vegan cookbook; it's a book about vegetables, and at times I use dairy. If needed, swap in your favorite plant-based dairy (see Cauliflower Bolognese, page 180) or vegan butter. I've listed alternatives wherever possible in the Cook's Notes for each recipe.

The Vegetable Pantry

It's not surprising that a lot of factors can change the taste and texture of our produce. If conditions are hospitable, microbes present in our food—such as bacteria, molds, and yeast—will grow and spoil your produce. For centuries, our ancestors developed various methods to make food last longer by storing grains and produce in dark containers in cool locations; using sugar, acids, and salt in food preservation; and even using fermentation. Let's look at some ways to store fresh produce.

What creates problems?

If we strip it down to a basic equation, the quality of our fresh produce can benefit from or be harmed by the vegetable's biology and the environment it's stored in, both affected by time.

Quality of Stored Fresh Produce = (Time) x (Plant Biology + Environment)

The goal here is to minimize or eliminate these factors to collectively prevent food spoilage and bacterial growth. Our pantries and appliances like refrigerators manipulate these factors to help our food last longer with minimal loss in quality.

Time

Time is perhaps the most obvious factor to affect quality, as the longer produce is stored, the more its quality diminishes. When vegetables deteriorate, they undergo biochemical changes that occur in their rates of respiration, transpiration, and metabolism. These change their aroma, taste, and texture. For example, over time green peas and sweet corn lose their sweetness as their sugars dissipate or turn into starch. Green leafy vegetables such as spinach and kale begin to wilt and turn yellow. Vegetables also lose water and start to dry up, as seen in the wrinkly and shriveled texture of elderly carrots.

The nutritional quality of vegetables also decreases with time as vitamins are lost. As the process of deterioration continues, bacteria and mold set in, and the vegetable rots. We do our best to stretch shelf life, if possible, by buying produce at its freshest, and controlling other factors like the biology of the fruit and vegetable and the storage environment.

Plant Biology

Every fresh vegetable (and fruit) has its own agenda. As soon as it's picked or enters its ripening phase, its cells kick off a series of biochemical reactions by breathing and releasing gases such as carbon dioxide and ethylene.

Respiration

Just like us, fresh produce breathes in oxygen and releases carbon dioxide into the air. The problem with your produce breathing is that the carbon dioxide arises from the breakdown of sugars and carbohydrates stored inside the vegetable or fruit. This will eventually cause both taste and texture to deteriorate, leading to spoilage.

Ethylene

Plants produce a sweet, musky gas called *ethylene* in response to stressful conditions such as low water or excessive heat and to aid ripening. Ethylene acts as a switch, triggering aging (senescence), the ripening of fruits and vegetables, and the yellowing (breaking down the green plant pigment, chlorophyll) and dropping (abscission) of leaves.

Fruits and vegetables like apples, bananas, tomatoes, and peppers (I know, I know; botanically speaking, these are all fruits) produce a lot of ethylene as they ripen. If leafy green vegetables like spinach are stored with these fruits, the leaves quickly turn yellow and decline, going limp and eventually turning to sludge. To avoid this and increase shelf life, store these ethylene producers separately from ethylene-sensitive produce in the refrigerator or pantry, or on the kitchen counter. Conversely, if you want to hasten ripening, you can expose produce that is sensitive to ethylene to a fruit that produces this hormone. For example, storing a banana next to a mango will help the mango ripen. Use the table on page 20 as a guide to storing your produce.

Enzymes

I once left a tomato on my kitchen counter and forgot about it for three or four weeks. You can imagine my shock when I cut through it to see the mealy flesh and seeds sprouting. No doubt you've had onions become mushy, cucumbers go soft, and sweet corn that tasted like starch after being left on the kitchen counter for a few days.

Once a fruit or a vegetable is harvested, it technically enters its next phase: deterioration. While we usually think of ripening increasing its

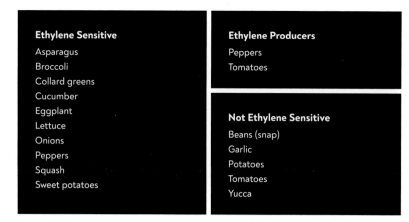

Ethylene Sensitive	**Ethylene Producers**
Asparagus	Peppers
Broccoli	Tomatoes
Collard greens	
Cucumber	**Not Ethylene Sensitive**
Eggplant	Beans (snap)
Lettuce	Garlic
Onions	Potatoes
Peppers	Tomatoes
Squash	Yucca
Sweet potatoes	

desirability, the ripening of produce is one of the early signs of deterioration. It's part of nature's process to ensure the propagation of the plant.

When some vegetables, like cucumbers and potatoes, are bruised or cut open, almost immediately their flesh becomes soft and brown. Physical damage to a vegetable breaks its cells, releasing enzymes that can destroy the surrounding vegetable tissue, and enzymes like polyphenol oxidase cause browning.

Environment
By manipulating the environment in which our produce is stored, we can minimize, slow down, and effectively eliminate the biological processes such as ripening and respiration that lead to spoilage. Exposure to light, unfavorable temperatures, humidity, and oxygen in the surrounding air can contribute to creating unfavorable conditions for vegetable shelf life.

Light
I rarely keep fresh produce (or even dried grains and cereals) exposed to sunlight or even kitchen lights. The exception to this is if I need to speed up ripening (for tomatoes, for example). For the most part, I store my produce away from sunlight. Vegetables such as endives start to turn green in the presence of light

and lose their flavor. Light also breaks down flavor molecules—like piperine, the substance responsible for the heat in black pepper—and renders the produce flavorless. Store produce such as potatoes, onions, and garlic and dried beans and spices in a cool, dark spot in your pantry or kitchen cabinets. Fresh vegetables that need colder temperatures should be kept in the refrigerator.

Temperature
According to the USDA, the ideal temperature storage conditions for most foods are as follows:

Dry foods	50°F [10°C]
Refrigerated foods	32°F to 40°F [0°C to 4.4°C]
Frozen foods	≤ 0°F [-18°C]

Not all fresh produce will freeze well; for example, bamboo shoots and jícama lose their crunchiness when their water content is frozen, then thawed. Because all vegetables are different, it is not possible to list a single temperature recommendation for their storage; however, there is a general rule that cool-season crops should be kept at 32°F to 35°F [0°C to 1.7°C] while warm-season crops at 40°F to 45°F [4.4°C to 7.2°C].

Humidity
Humidity is simply the amount of water that's vaporized in the air. Refrigerators circulate cool air to keep temperatures low, but this can cause food to dry out. To avoid this, maintain your refrigerator's humidity between 60 and 75 percent at 55°F [13°C]. The crisper drawers in your refrigerator are specifically designed to provide higher humidity for vegetables and lower for fruits. Many refrigerators now let you control the amount of humidity, and some use smart technology to sense and adjust humidity as needed.

Oxygen
Oxygen tends to reduce the quality of food because it leads to increased ethylene and carbon dioxide production. Commercial food storage companies usually get rid of the air that would surround food products by using vacuums or by replacing the air with a layer of nitrogen or carbon dioxide (so the vegetables can't breathe). With some vegetables, like garlic and onions, it's best to leave their papery wrappers intact so they last longer. These days, most refrigerators come with built-in air filters that remove a lot of the gases that lead to unpleasant odors and food spoilage.

Adapted from Post Harvest Center, University of California, Davis.

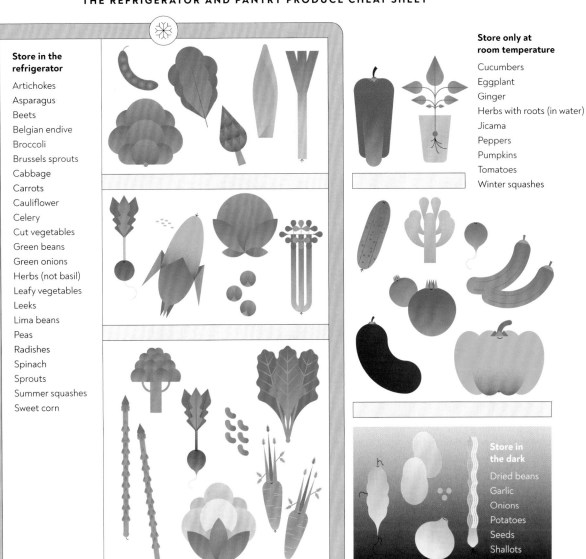

Store in the refrigerator

Artichokes
Asparagus
Beets
Belgian endive
Broccoli
Brussels sprouts
Cabbage
Carrots
Cauliflower
Celery
Cut vegetables
Green beans
Green onions
Herbs (not basil)
Leafy vegetables
Leeks
Lima beans
Peas
Radishes
Spinach
Sprouts
Summer squashes
Sweet corn

Store only at room temperature

Cucumbers
Eggplant
Ginger
Herbs with roots (in water)
Jícama
Peppers
Pumpkins
Tomatoes
Winter squashes

Store in the dark

Dried beans
Garlic
Onions
Potatoes
Seeds
Shallots

Refrigerator Help

One way to store both fresh and cooked produce is to cut access to air by using ziptop bags or containers that can be sealed. Food vacuum devices (I love my Zwilling Fresh & Save vacuum) are also fantastic at getting rid of air but be careful when using them with raw vegetables like onions, garlic, and even mushrooms; these can carry harmful bacteria from the soil that can survive in the absence of oxygen. These vegetables should be stored in a breathable environment; for example, mushrooms can be stored in a paper bag in the refrigerator.

To store fresh herbs and salad leaves, I like to wrap them in paper towels before washing, as the towels help wick away moisture and prevent the leaves from becoming slimy. Some advanced food storage systems use ethylene absorber kits built into storage containers; these absorb moisture for extra insurance. Save those silica gel sachets that come packaged with dried foods like seaweed snacks; you can reuse them when storing dried produce such as seeds. You can store certain cut vegetables, like carrots, bell peppers, and bamboo shoots, in fresh water in the refrigerator; just remember to change the water daily.

29

Onions
Shallots
Scallions
Leeks
Garlic
+ Chives

The Amaryllis Family

AMARYLLIDACEAE

They can make the mightiest warriors weep and drive vampires away. Could there be a more powerful vegetable family on our planet?

Origins

THE GEOGRAPHIC ORIGIN OF ALLIUMS IS UNCERTAIN, BUT MOST RESEARCH POINTS TO ASIA. FOR SHALLOTS, IT'S CENTRAL ASIA. FOR SCALLIONS, ASIA. FOR LEEKS, THE MEDITERRANEAN AND THE MIDDLE EAST. FOR GARLIC, THE PATH LEADS FROM CHINA TO INDIA TO EGYPT TO UKRAINE. FOR CHIVES, EUROPE, ASIA, AND POSSIBLY NORTH AMERICA.

Onions

I have a personal relationship with onions: If it weren't for them, I might have attended culinary school. When I vacillated between a career in science and a career in cooking, my mother steered me toward science. Her reasoning was that I didn't have the stamina to peel and prep onions in a cold room. She'd seen many young chefs who staged at the hotel she worked at, and many of them were tasked with prepping large bags of onions in the walk-in refrigerator rooms where vegetables were stored (the colder temperatures reduced the risk of crying). Luckily, life had other plans for me, and I eventually did become a cook, despite the onions.

Onions form the basis of many sauces, stews, curries, marinades, and even garnishes. Crispy fried onions are a delight on top of biryanis and pilafs, as are onion rings when breaded and fried (see Golden Za'atar Onion Rings, page 36). I prefer to use red onions in raw dishes, like salads, while white and yellow onions I use raw but also for cooking. Buy onions that still have their dry papery skins attached, and avoid ones that are sprouting.

Shallots

Shallots are sweeter than onions (they contain almost three times as much natural sugar as onions) but that doesn't mean they won't make you cry when you cut them. Shallots are lovely both raw and cooked. I often prefer them to onions, especially when I need a small quantity of onion flavor in a dish (see Chaat-Style Loaded Twice-Baked Potatoes, page 291). Because shallots contain so much sugar, I find they're better than onions for making onion jam.

Scallions

Imagine a young onion in an herb form but bolder than chives. Scallions, or green onions, are one of my favorite garnishes, and I use both the white and the green parts. They're also great sautéed and in stir-fries. In the United States, the term *spring onion* refers to the young onions with bulbs larger than a scallion's and with green stalks still attached. In the United Kingdom and many other parts of the world, spring onion is also used to describe a scallion.

Leeks

In food writing, there are two ingredients that are always described as "melting" during cooking: anchovies and leeks. Neither really "melt," but when cooked the heat transforms them into something glorious, with the softest texture that simply gives way with each bite.

Leeks are like a big scallion; you trim the ends off and use the white and green parts. Some grocery stores trim their leeks; avoid buying those and instead buy leeks that are intact. Leeks tend to accumulate a lot of sand and grit as they grow, so make sure you rinse them extremely well before you cook with them. I once eagerly bit into a leek omelet (where that was I will not reveal) only to encounter the grit of poorly washed leeks. It felt like eating sandpaper. I rinse leeks a few times before I cut, and after slicing, I rinse and let them sit in a bowl of cold water. Any dirt that's trapped will sink to the bottom of the bowl, and the sliced leeks, lightened, will float to the surface, where you can carefully remove them and leave the sediment behind.

Garlic

Known for its ability to ward off vampires and destroy date nights, garlic is a powerful ingredient. But it's also a delicious and essential ingredient, so it remains justly popular. I use garlic liberally when I cook. How garlic is prepped does make a difference in its flavor intensity. Grated garlic tends to taste much stronger when used raw, but when cooked most of those aromatic flavors quickly mellow. In comparison, sliced and minced garlic tend to taste much stronger when cooked. There are two main types of garlic: the ones with a woody stalk, called *hardneck*, and the other, *softneck*, which is the variety most often seen in grocery stores. Softneck tends to be slightly milder in flavor than hardneck.

Most of us are used to garlic heads that contain several cloves; however, the pearl variety contains a single large clove. The long, curvy gooseneck-shaped stalk or scape (a flower stalk) that hardnecks form is another popular part of fresh garlic that's used in cooking. It is often sautéed in olive oil or incorporated into pesto. In this book, a clove of garlic is one large clove unless otherwise stated.

Garlic powder or ground dried garlic is a great option in recipes when you want to limit the liquid content to, for example, help veggie burgers hold their structure (see Masala Veggie Burgers, page 330). Garlic powder tends to absorb moisture over time and harden, so store it in an airtight container (this is another place to add a desiccant packet). If your garlic powder has solidified, try to cut the powder with a sturdy knife and then pound it with a mortar and pestle before using.

Black garlic is a beautiful form of garlic that originated in Asia; it tastes sweet with hints of caramel. This special form of garlic, an ingredient prized by many chefs, is produced by either leaving garlic to ferment in a cool, humid place or by slowly cooking it at a very low temperature using a slow cooker or sous vide device. Some brands of black garlic are made with a single large clove of pearl garlic (keep this in mind when estimating how much to use, as you will need fewer of these large cloves). For all practical purposes, I use a whole head (which is one clove) of black pearl garlic for an entire head of black garlic. During production the pungency of the garlic dissipates, the sweetness increases, and caramelization and the Maillard reaction produce deep notes of caramel and bittersweet flavors. I keep a big jar of black garlic on hand to fold into dairy like yogurt or crème fraîche (see the crème fraîche dip with the Bombay Potato Croquettes,

page 266), vegetables, and even pasta (add 2 Tbsp of smashed black garlic cloves to the Shallot + Spicy Mushroom Pasta, page 45). Because black garlic is very soft and sticky, the best ways to prepare it for use are to either pulverize it into a pulp in a food processor or blender, or smash it into a paste with the blunt side of your knife on a cutting board.

Another member of the allium family, elephant garlic, is not a true garlic but rather a type of leek. It has a mild flavor, and I don't recommend using it as a garlic substitute.

Chives

We treat chives more as an herb and less as a vegetable, folding them into dips and sprinkling them on top of salads, rather than centering a dish around them, as we do with braised leeks or studded onions. There are three main types of chives: the ones that taste like onions, called *common chives* or *onion chives*; another oniony variety called *blue chives* or *Siberian chives*; and garlicky *Chinese chives* or *garlic chives*. Chives make a fabulous garnish and I use them often as a finishing touch in many dishes, such as the Leek + Mushroom Toast (page 42) and in pasta (Shallot + Spicy Mushroom Pasta, page 45). The flavor molecules in chives are soluble in fats, and the Sichuan butter (Corn Cakes, page 49) uses this property to create an extremely flavorful infused butter.

Storage

Store alliums such as onions, shallots, and garlic in a cool, dark place away from sunlight. Keep them in a wire basket or a paper bag to let them breathe; this reduces the buildup of moisture and decreases the risk of mold. Tender fresh alliums such as scallions, leeks, and chives should be stored in the refrigerator because

they will dry out in a pantry. I'm not going to lie: Frozen garlic (and ginger) is one of the best processed food items available in grocery markets because it is as good frozen as fresh and great to have on hand in case of a seasoning emergency.

Cooking Tips

Cut a raw onion and it will make you weep. Eat raw garlic and your date might end up a disaster. *Alliums* are famous for their pungency. When an allium is cut, the cells break and release an enzyme called *alliinase* that produces volatile chemicals. In the case of onions and shallots, this enzyme makes our eyes sting and water; in the case of garlic, it leaves that lingering aroma. The enzyme alliinase produces its most potent chemicals at warmer temperatures; cold temperatures will reduce the activity of this enzyme. So to minimize "crying" when chopping onions and shallots, I stick the onions and shallots in the fridge for an hour or two before cutting them. (If I haven't planned ahead, I wear goggles!) To reduce the pungent aroma of raw onions and shallots in salads, I soak them in a bowl of ice water for 20 minutes, then drain and pat them dry before adding them to the salad.

Alliums are rich in sugars and contain long chains of the sugar fructose (fructose is also found in honey) called *fructans*. There are a couple of different ways to bring out their sweetness, and almost all of them rely on destroying the enzyme alliinase and breaking the fructans chains. Onions and garlic pickled in an acid like vinegar or lemon juice are sweet to taste. The low pH provided by the acid inactivates the enzyme alliinase, and the taste of the sugars becomes obvious. The acid also helps break down some of the chains of fructose and, in turn, increases sweetness. When heat is applied

33

using cooking methods—roasting, frying, sautéing, and sous vide—the higher temperatures destroy the alliinase enzyme, help break down the long sugar chains, and heighten the sweet taste.

If there is one technique that comes up again and again in this book—and, if I may be so bold as to say, in the world—it is the frying and sautéing of alliums. Not only do frying and sautéing help increase sweetness and eliminate raw pungency, but the fat acts as a powerful solvent to draw out the flavor chemicals from alliums. The aromatic chemicals that give alliums like garlic and onions their flavor dissolve in fat with great joy. The fat holds onto those flavors and becomes extra tasty. The two major food browning reactions, caramelization and Maillard, help create these bittersweet flavors and various shades of caramel during cooking.

There are several recipes in this book (see Shallot + Spicy Mushroom Pasta, page 45) that rely on the browning of alliums to maximize flavor. Because there are so many variables involved with browning—the alliums, the heat source, the cooking utensils, as well as our own personal preference of how far to take foods we're browning—it would be foolish for me to list a cooking time. For this reason, I've skipped listing "browning time" estimates whenever onions and shallots are caramelized.

During browning, onions and shallots inevitably dry out and can easily burn. To avoid this, I use a tip I learned from food writer Ali Slagle: Add 1 or 2 Tbsp of water. The steam generated will help reduce the cooking temperature and prevent drying.

Quick Pickled Red Onions or Shallots

This general recipe for Quick Pickled Red Onions or Shallots goes great in almost any dish, but especially over a pot of dal, in Lentil Lasagna (page 242) or Goan Pea Curry (page 235), or on the Masala Veggie Burgers (page 330).

MAKES 4 SERVINGS

In a jar with a lid, combine **1 cup [140 g] thinly sliced red onion or shallots**, **½ cup [120 ml] apple cider vinegar**, **2 Tbsp chopped fresh herbs (dill, chives, tarragon, cilantro, or parsley)**, **1 tsp toasted spices (cumin, coriander, fennel, or nigella)**, **½ tsp sugar**, and **⅛ tsp fine sea salt**. Refrigerate for at least 30 minutes before using. Store refrigerated for 1 to 2 days.

Golden Za'atar Onion Rings with Buttermilk Caraway Dipping Sauce

MAKES 4 TO 6 SERVINGS

To prepare the onion rings, toss **2 extra-large white or yellow onions, sliced into ½ in [13 mm] thick rings,** separated, with **1 tsp fine sea salt** in a large mixing bowl. Cover and let sit for 1 hour at room temperature.

Buttermilk Caraway Dipping Sauce

Meanwhile, prepare the buttermilk caraway dipping sauce. In a blender or a food processor, blend **1 cup [240 ml] plain unsweetened buttermilk or kefir**; **½ cup [20 g] chopped cilantro**; **½ cup [5 g] chopped dill**; **2 garlic cloves, roughly chopped**; **1 green chilli such as jalapeño, stemmed and roughly chopped**; and **1 tsp whole caraway seeds** until smooth, 30 seconds to 1 minute. Taste and season with **fine sea salt**. Transfer to a small serving bowl, cover, and keep chilled until ready to serve. The sauce can be made a day ahead of time and stored in an airtight container in the refrigerator until ready to serve, up to 3 days.

Put the onions into a fine mesh sieve, let the liquid run off, and rinse with cold water. Pat dry with a lint-free kitchen towel or paper towels.

When ready to fry, line a baking sheet with a layer of absorbent paper towels or a wire rack.

In a large mixing bowl, whisk together **2 cups [280 g] all-purpose flour, ½ cup [70 g] cornstarch, 1 tsp ground black pepper, 1 tsp fine sea salt,** and **½ tsp ground turmeric**.

In a second large bowl, whisk until combined **2 cups [480 ml] buttermilk or kefir, ½ tsp fine sea salt,** and **½ tsp ground turmeric**.

Work with a quarter of the onions at a time to avoid overcrowding. Set up an assembly line with the onions, your two bowls of coating mixtures, and a large unlined baking sheet or tray. Using a pair of kitchen tongs or two forks, add the onions to the flour mixture, toss to coat well, and tap the onion rings against the side of the bowl to remove excess flour. Dip the rings in the buttermilk mixture, toss to coat well, then tap them on the side of the bowl to get rid of any excess liquid. Return them to the bowl

36

with the flour, toss to coat well, and again tap to remove excess flour. Place the coated rings on the baking sheet.

In a large, deep frying pan, warm over medium heat **3 to 4 cups [710 to 945 ml] neutral oil** with a high smoke point, such as grapeseed, to 350°F [180°C].

Fry the onions in the hot oil, stirring to separate the rings, until they are golden brown and crisp, 6 to 7 minutes. Transfer the onions to the prepared baking sheet. In batches, fry the remaining onions, letting the temperature return to 350°F [180°C] before adding the next batch. Toss the hot, fried onion rings with **2 Tbsp za'atar, homemade (page 341) or store-bought**.

Serve immediately with the buttermilk caraway dipping sauce on the side.

I love fried food and herby dips, but there's a more profound message embedded here: The combination of hot and cold temperatures is a joyful experience. The concept of contrasting temperatures works elegantly when hot, crispy onion rings kissed with turmeric and za'atar are dunked into the cold herby buttermilk dipping sauce. This makes a worthy appetizer; these rings are also great stuffed into a bun with the Masala Veggie Burgers (page 330) and the buttermilk caraway sauce. Serve with chilled ginger ale or beer.

THE COOK'S NOTES

- Salting the onions helps draw out their moisture through osmosis, softening the cells' tough pectin just enough (do not soak for more than 1 hour), creating a crisper texture and more uniform taste.

- Avoid dredging the onions using your hands; it is messier, it leads to an uneven coating on the fried onions, and you might end up with not enough dredging mixture (most of it ends up sticking to your fingers). Two forks or a pair of kitchen tongs are your friends here.

- For an extra-crispy texture, whisk 2 Tbsp fine semolina into the flour.

- Buttermilk and kefir are interchangeable in this recipe; they are both tangy and creamy, and help bind the breading mixture to the onions. If using kefir, opt for a brand that isn't particularly thick (Lifeway and Green Valley Creamery are two of my favorites).

Saffron Lemon Confit with Alliums + Tomatoes

MAKES 4 TO 6 SERVINGS

Preheat the oven to 350°F [180°C]. Heat **2 Tbsp extra-virgin olive oil** over medium heat in a large cast-iron or stainless-steel skillet. Add **2 large leeks, white and light green parts only, trimmed and cut crosswise into thin slices**, and **4 shallots, cut in half lengthwise**. Cook, flipping the leeks over with a pair of kitchen tongs, until fragrant, the leeks have softened, and the shallots are lightly browned, 3 to 4 minutes. Add **1 small head garlic, halved through the equator**, cut side down, adding more oil to the skillet if necessary. Cook until the garlic is fragrant and just starting to brown, about 1 minute. Remove from the heat and transfer the vegetables to an 8 by 6 by 2 in [20 by 15 by 5 cm] deep baking dish or loaf pan.

Add **1 lemon, cut into thin slices (seeds discarded)**, and **1 pint [280 g] cherry or grape tomatoes, each pricked with a skewer or fork**. Sprinkle with **1 tsp red pepper flakes such as Aleppo, Maras, or Urfa** and **a large pinch (15 to 20 strands) saffron**. Pour **1 cup [240 ml] extra-virgin olive oil** over the vegetables in the dish (add more if needed to cover).

Cover the dish with aluminum foil and seal tightly all around. Place on a rimmed sheet pan (in case it bubbles over) and bake for about 1 hour. The leeks and garlic should be falling apart, and the shallots should be soft when done. Remove from the oven and let sit, covered, for 10 minutes before serving.

Sprinkle with **a little flaky salt** and serve warm with **bread**. Save leftover confit in the refrigerator. Warm to room temperature before serving as a dipping oil with bread. You can also make this a week in advance and store in an airtight container in the refrigerator. Warm it up in the oven at 300°F [150°C].

When it comes to slow cooking, confit (derived from the French word *confire*, "preserve" or "crystallize") is perhaps one of the most versatile techniques. A confit is prepared by submerging raw vegetables or meat in a liquid that prevents the growth of harmful microbes—oil or concentrated sugar syrup, for example—and slow-cooked for a long time at low heat. In this instance, olive oil is used as the medium in which garlic, leeks, shallots, and tomatoes are slow-cooked. Unlike deep-frying, this method starts out with oil at room temperature. It warms up slowly in the oven, resulting in a more concentrated flavor and tender texture.

THE COOK'S NOTES

- To maximize the flavor of the vegetables, the alliums are first lightly browned in oil. The combination of caramelization and the Maillard reaction helps create a wondrous array of brown colors and bittersweet flavors.

- The bright orange color and musky flavor of saffron are fat-soluble and permeate through the olive oil during cooking.

- Pricking the tomatoes helps release their liquids into the oil to create a more concentrated flavor.

- You need a small baking dish to submerge this amount of vegetables in the olive oil. If your cooking vessel is too wide, they won't be covered, and they won't cook properly. If all you have is a large baking dish, then add enough olive oil to completely cover the vegetables. A 9 in [23 cm] loaf pan will also work here.

1. Onions, Shallots, Scallions, Leeks, Garlic + Chives

Red Onion + Tomato Yogurt

In a large serving bowl, combine **2 cups [480 g] plain, unsweetened full-fat Greek yogurt or labneh; 1 garlic clove, grated**; and **1 Tbsp fresh lemon juice**. Season with **fine sea salt** and **½ tsp ground black pepper**.

Heat **2 Tbsp extra-virgin olive oil** in a small cast-iron or stainless-steel skillet over medium-high heat. Add **1 small red onion, halved lengthwise and cut into thin crescents**. Fry until the onions start to brown, stirring often, about 6 to 8 minutes.

Add **1 pint [280 g] cherry or grape tomatoes**. Sauté until the tomatoes just start to burst, 2 to 3 minutes. Add **a pinch of fine sea salt**. Remove from the heat and top the yogurt with the onions and tomatoes.

Prepare the tadka. Wipe the same skillet clean with a clean paper towel and warm **2 Tbsp extra-virgin olive oil** over medium heat. When the oil is hot, add **1 tsp whole nigella or cumin seeds** and **½ tsp ground coriander**. Fry the seeds, stirring constantly, until fragrant and they start to turn light brown, 30 to 45 seconds. Remove from the heat and add **½ tsp red pepper flakes such as Aleppo or Maras**. Swirl the oil in the skillet and pour the mixture over the onions and tomatoes.

Garnish with **1 Tbsp chopped fresh oregano** and **1 Tbsp chopped fresh basil**. Sprinkle with **a little flaky salt (optional)**. Serve immediately.

In our house, I cook almost everything, but when it comes to grilling, that job goes to my husband. I like my accompanying dishes to stand on equal footing next to the main dishes, and this simple yet elegant dish of sweet, browned onions and tomatoes does just that next to his grilled steaks and pork chops. The Indian flavoring technique of tadka, where spices are infused in hot oil, transforms what could be a supporting player into a breathtaking star of the meal. I sometimes make this for breakfast and add a few poached eggs on top (like çilbir, or Turkish eggs) and serve it with buttered slices of toasted sourdough bread.

THE COOK'S NOTES

- Both Greek yogurt and labneh will work here. Some brands are saltier than others, so taste the labneh before you season it to get a sense of how salty it is.

- Nigella and cumin seeds don't taste the same, but separately they work brilliantly with the yogurt. To get a full appreciation of what each brings to this dish, just use one or the other, not a combo.

- When preparing the tadka, listen to it carefully. The frying spices will sing—crackling and popping—a cue that tells you whether the oil is hot enough and also when it's ready to pull off the stove.

Leek + Mushroom Toast

MAKES 4 SERVINGS

Melt **2 Tbsp unsalted butter** in a large stainless-steel skillet over medium-high heat. Cook until the milk solids turn golden brown and the water in the butter evaporates and stops crackling, 2½ to 3 minutes, partially covering the pan as needed to protect from splattering.

Add **2 Tbsp extra-virgin olive oil** and fry **½ tsp ground cumin** and **½ tsp ground black pepper** until fragrant, 30 to 45 seconds, stirring constantly to avoid scorching. Turn down the heat to medium, add **2 large leeks, white and light green parts only, trimmed and thinly sliced**, and sauté until lightly golden brown, 7 to 10 minutes. Fold in **8 oz [230 g] cremini mushrooms, trimmed and thinly sliced**, and sauté until they start to caramelize and turn reddish-brown, 3 to 4 minutes.

Add **2 garlic cloves, minced**, and **2 tsp white or yellow miso** and sauté until fragrant, 30 to 45 seconds.

Stir in **1 cup [240 ml] water** and **1 tsp low-sodium soy sauce** to blend in the miso. Increase the heat to high and bring to a boil. Lower the heat to a simmer and cook, uncovered, stirring occasionally, until the liquid reduces by three-quarters and thickens, 12 to 15 minutes. Taste and season with **fine sea salt**.

While the mushroom mixture cooks, prepare the toast. Preheat the oven to 350°F [180°C] and set a rack in the middle.

Cut **4 slices of good-quality sourdough bread** each about ½ in [13 mm] thick. Brush each slice on one side with a little extra-virgin olive oil or softened butter. Place them directly on the middle oven rack and toast for 6 to 8 minutes, until lightly golden brown and crispy. (You can also use your toaster and brush with a bit of oil after toasting.)

Divide the mushroom topping among the 4 toasts. Garnish with **1 Tbsp chopped chives**, divided evenly among the 4 toasts. Don't wait; serve immediately. No one likes soggy toast unless it's bread pudding. If you have leftover topping, store, refrigerated, in an airtight container for up to 3 days.

"Loaded minimalism" sounds paradoxical, but that's the only appropriate way to describe how I like my toasts. Few ingredients—but plenty of them. Give me a loaded toast topped with leeks that are tender enough that they're falling apart and succulent slices of mushrooms, and I'm a happy camper. The combination of miso and garlic amps up the savory profile of this toast.

THE COOK'S NOTES

- You really don't need to stick to cremini mushrooms; this recipe is quite amenable to other kinds. The cooking time will change a little, and with smaller types of mushrooms like enoki, skip cutting them but do separate them.

- This is a reminder to wash leeks very well. It's an awful feeling to take a big mouthful and bite down on sandy grit.

1. Onions, Shallots, Scallions, Leeks, Garlic + Chives

Shallot + Spicy Mushroom Pasta

MAKES 4 SERVINGS

Melt **2 Tbsp unsalted butter** in a large 12 in [30.5 cm] deep-sided cast-iron or stainless-steel skillet over medium-high heat. Cook until the milk solids turn golden brown and the water in the butter evaporates and stops crackling, 2½ to 3 minutes, partially covering the pan as needed to protect from splattering.

Add **2 Tbsp extra-virgin olive oil** and sauté **8 shallots, cut in half and thinly sliced (total weight about 15¼ oz [435 g])**, and **½ tsp fine sea salt** until they turn a dark golden brown, 25 to 30 minutes. Stir often to prevent burning. Be warned: Browning onions and shallots is one of the most controversial things in recipe writing; predicting a definitive time for them is a fool's errand. If they start to burn, use recipe writer Ali Slagle's trick and add **1 to 2 Tbsp water** to lower the temperature of the pan and continue to cook.

Add and sauté **4 garlic cloves, minced**; **1 tsp red pepper flakes such as Aleppo, Maras, or Urfa**; and **¼ tsp dried sage**. Cook until fragrant, 30 to 45 seconds.

Add and sauté **8 oz [227 g] cremini mushrooms, quartered**, until they turn brown and release their liquid, 4 to 5 minutes. Stir in **zest of 1 lemon** and **1 Tbsp fresh lemon juice**. Taste and season with **fine sea salt**.

Bring a large pot of salted water to a boil and cook **1 lb [455 g] spaghetti** until al dente, per the package instructions. Reserve **1 cup [240 ml] of the cooking water** and drain the cooked pasta.

Add the pasta to the pan and toss with the shallot-mushroom mixture. Add **½ cup [120 ml] of the reserved cooking water**. Fold to coat well, adding more pasta water as needed. Taste and season with **fine salt** and **pepper**. Garnish with **¼ cup [15 g] grated Parmesan** and **2 Tbsp chopped chives**. Serve warm.

Store and refrigerate leftovers in an airtight container for up to 3 days.

continued

This recipe is a milestone: It's the first time I've included a pasta recipe in a cookbook. There are some cooking aromas I dream about, and the scent of caramelized shallots is one of them. I think of shallots as tinier and sweeter versions of red onions. Once caramelized, they declare their savory sweetness proudly, a perfect accompaniment to the mushrooms. There's a tip that I've borrowed from Indian biryanis: use plenty—and I do mean plenty—of shallots for a magnificent effect.

THE COOK'S NOTES

- I'm going to give a tip on what *not* to do: Do not add baking soda to accelerate caramelization and the Maillard reaction with shallots or red onions. Shallots and red onions get their pink color from the food pigment anthocyanin, which is sensitive to changes in pH. Adding baking soda will raise the pH and can turn the pink pigment a hideous shade of green and release a lot of water (baking soda breaks down the pectin in the cell walls of the shallots), making a greenish mushy mess of your beautiful shallots.

- When chestnuts are in season, I'll slice thin sections of roasted chestnuts and sauté them with the mushrooms.

- Fried or grilled strips of halloumi make a fantastic alternative to Parmesan.

Corn Cakes with Sichuan Chive Butter

MAKES 4 SERVINGS

Sichuan Chive Butter

To prepare the butter, in a small bowl fold together **½ cup [110 g] unsalted butter, softened**; **3 Tbsp chopped chives**; **2 generous Tbsp solids from a jar of chili crisp**; and **1 garlic clove, grated**. Add, as needed, **flaky sea salt**. The butter will store well for up to 3 days in an airtight container in the refrigerator.

Prepare the corn cakes by whisking together **1 cup [140 g] medium-grind cornmeal, 1 cup [120 g] cake flour, 1 tsp fine sea salt, ½ tsp ground turmeric, ½ tsp baking powder,** and **¼ tsp baking soda**.

In a medium mixing bowl, whisk together **1 cup [240 ml] buttermilk or kefir, ¼ cup [50 g] packed dark brown sugar, 2 large eggs,** and **2 Tbsp melted ghee, butter, or other neutral oil with a high smoke point such as grapeseed**. Make a well in the center of the cornmeal mixture, pour in the liquids, and whisk until smooth and combined.

Melt **½ Tbsp unsalted butter** in a large cast-iron skillet or on a griddle over medium-low heat. Spoon out one-fourth of the batter in the center and spread evenly into a circle that's approximately 6 in [15 cm] round. Cover the skillet with a lid and cook until the corn cake starts to become crispy and golden brown on the bottom, 2 to 3 minutes. Then flip the corn cake with a large spatula, cover, and cook again until crispy and golden brown on the other side, 2 to 3 minutes. Transfer to a plate and cover with a kitchen towel to keep warm. Prepare the remaining corn cakes, adding more butter to the pan as needed between batches.

Serve the corn cakes warm topped with a generous pat of the Sichuan chive butter. Store leftover cakes refrigerated, in an airtight container; store the butter wrapped in parchment paper in an airtight container.

Some people love pancakes for breakfast; I love corn cakes. I love them so much, I'd run a marathon if they were waiting for me at the finish line. These golden corn cakes with their crispy brown edges, topped with a generous spoonful of melting Sichuan chive butter, are the perfect combination of sweet, savory, spicy, and hot. Be generous with the butter.

THE COOK'S NOTES

- Cake flour has a lower protein content, so it produces a much more tender crumb than all-purpose flour.

49

Roasted Garlic + Chickpea Soup

MAKES 4 SERVINGS

Preheat the oven to 400°F [200°C].

Coat **1 small head garlic, outer papery layers removed and top ¼ in [6 mm] trimmed**, with **½ Tbsp extra-virgin olive oil**. Wrap the garlic in aluminum foil and roast for about an hour, checking after 45 minutes. Allow the garlic to cool in the foil until cool enough to handle. Squeeze the garlic cloves from their husk. Transfer to a blender.

Drain and rinse **two 14 oz [400 g] cans of chickpeas**. Set aside one-quarter of the drained chickpeas. Place them on a kitchen towel, pat dry, and set aside. Add the rest to the blender.

Add to the blender **3 cups [710 ml] low-sodium vegetable or beef stock or Master Mushroom Vegetable Stock (page 337)**; **2 Tbsp apple cider vinegar**; **2 Tbsp white or yellow miso**; **½ tsp ground turmeric**; **½ tsp ground black pepper**; **½ tsp ground cinnamon**; and **¼ tsp ground cayenne**. Blend on high speed until smooth. Pour the liquid into a medium saucepan and bring to a boil. Taste and season with **fine sea salt** and **vinegar**. Keep warm over low heat if not serving immediately.

Heat **2 Tbsp extra-virgin olive oil** in a small saucepan over medium-high heat. Fry the reserved chickpeas until they turn crisp and golden brown, 4 to 5 minutes. Turn down the heat to low, sprinkle with **½ tsp dried oregano or 1 tsp chopped fresh oregano**, **½ tsp ground cumin**, **½ tsp sweet smoked paprika**, and **¼ tsp ground cayenne**. Taste and season with **fine sea salt**. Swirl to coat well and fry until fragrant, 30 to 45 seconds. Remove from the heat.

To serve, divide the warm soup among four bowls. Garnish each with the crispy chickpeas and drizzle with a little extra-virgin olive oil.

Roasting garlic may be my favorite of all the transformations that take place in the kitchen. Garlic's pungency mellows, turning into sweet and caramel-like notes. This is one of my favorite soups not only for its flavor, but also because it's quick to prepare (especially if you've got cooked chickpeas on hand). The fried chickpeas do a great job masquerading as croutons and bring a crunchy contrast to this splendid creamy soup.

THE COOK'S NOTES

- You can also use the crispy spiced chickpeas (page 225) in place of the fried chickpeas. Croutons are also wonderful here.

- The garlic can be roasted a day or two ahead of time.

- Besides paprika, other chillies with a smoky profile—chipotle and Kashmiri chilli powder—will also work.

Yams

The Yam Family
DIOSCOREACEAE

Origin
YAMS ORIGINATED IN AFRICA, ASIA, AND THE CARIBBEAN.

Yams

Yams, or nyami (the West African name), are often confused with sweet potatoes. Except for growing buried under the soil, these tubers aren't similar at all. Yet they are, for the most part, interchangeable in recipes.

Yams and sweet potatoes look different (Sweet Potatoes, page 133). The outer skin of a yam is rough and brown or black with hairy fibrous roots; the skin of a sweet potato is much smoother in comparison. Unlike sweet potatoes, yams aren't sweet, and I find their neutral flavor a fantastic palette for various spices and condiments. The texture of a cooked yam is more like a cross between a potato and a cassava: starchy, dryish, and fibrous.

If they look and taste dramatically different, you might be wondering where the confusion between yams and sweet potatoes arose. As author Margaret Eby wrote in *Food and Wine*, the confusion between these two vegetables arose during the transatlantic slave trade. Slave ships would carry captive people from Africa along with yams to feed them. But yams weren't grown or available in the Americas, and sweet potatoes—native to Central America and brought north into the United States—were available. Sweet potatoes took the place of yams in the diet of the enslaved people, the word *yam* deriving from various West African words for "to eat."

In the 1930s, the Louisiana Agricultural Experiment Station released a new variety of sweet potatoes developed by Julian C. Miller that boasted "an attractive skin, and a moist, orange flesh high in vitamin A content." Louisiana producers competed with the traditional drier, white-fleshed sweet potatoes. An aggressive marketing campaign followed and, in 1937, the Louisiana industry dubbed their product as "yam" in a calculated effort to make their orange varieties stand out. The rest, as they say, is history. The name *yam* stuck, and grocery stores across America called sweet potatoes "yams." When I refer to yams throughout this book, I am not referring to sweet potatoes; I am referring to the real-deal, native-to-Africa tuber.

Storage

Raw yams should be stored in a cool, dark spot in the kitchen and will keep for a couple of months.

Cooking Tips

- You won't find true yams in a regular grocery store, but they can be procured at African and Caribbean grocery markets or online. In Los Angeles, you can find them at Stone's Grocery and Market, where I buy the Ghanian puna yam. I find its taste and texture more like potatoes than sweet potatoes. The flavor reminds me of a nutty potato and the texture is slightly drier than a potato.

- Yams tend to be quite large; you can cut off and cook as much as you need. The cut part of the remaining vegetable will dry and protect the vegetable for a few weeks. The other alternative is to cook it all and use what you need, refrigerate the rest, and use it within a week.

- Ube, the bright purple yam that's extensively used in Filipino cuisine, is sometimes erroneously referred to as sweet potatoes or taro, which also come in purple varieties. Once cooked, the starch tuber should be mashed or puréed in a food processor to break up any fibers. It's usually then incorporated into desserts such as ice cream or cakes.

- Yams must be cooked thoroughly before eating. The heat helps destroy the naturally occurring toxic chemicals diosbulbins, histamines, and cyanogens, rendering them safe to eat.

- Raw yams turn slippery when peeled. I find that rinsing them under running water and occasionally patting them dry helps with stability during prepping. Wear gloves while working with them, as the sap of the vegetable might irritate your skin (just like chayote; see Cooking Tips, page 187).

- Peel yams prior to cooking. Remove all the skin completely, including the thinner brown layer and dark spots, or else they will turn black when heated, especially during roasting.

- I prefer boiling yams in a pot of salted water and then roasting them at 425°F [220°C] (see Sweet + Sour Yams, page 60), as the texture is much creamier than if they were simply roasted.

Mashed Yams with Tomato Sauce

MAKES 4 SERVINGS AS A SIDE

Place **1 lb [455 g] yams, peeled and cut into ½ in [13 mm] cubes**, and **1 tsp fine sea salt** in a medium saucepan filled with enough water to cover them by 1 in [2.5 cm] and bring to a boil over high heat. Lower the heat to a simmer, cover, and cook until the yams are tender and easily pierced by a fork or knife, 12 to 15 minutes. Remove from the heat, drain, and discard the water.

While the yams cook, prepare the sauce. In a medium saucepan over medium heat, warm **2 Tbsp extra-virgin olive oil**. Add **1 medium white or yellow onion, diced**. Sauté until it becomes translucent, 4 to 5 minutes. Add **5 garlic cloves, grated**; **1 tsp smoked sweet paprika**; **1 tsp ground cumin**; **1 tsp ground coriander**; **1 tsp ground sumac**; and **½ tsp ground cinnamon**. Sauté until fragrant, 30 to 45 seconds. Stir in one **14 oz [400 g] can crushed tomatoes**. Increase the heat to medium-high and bring to a boil, being careful to avoid splattering. Remove from the heat, taste, and season with **fine sea salt** and **a pinch of sugar** if needed.

Transfer the cooked yams to a large mixing bowl. Mash with a fork or potato masher until smooth. Mix in **¼ cup [60 ml] extra-virgin olive oil**; **2 to 4 garlic cloves, grated** (see the Cook's Notes); and **½ tsp ground black pepper**. Taste and season with **fine sea salt**.

Transfer the yams to a serving bowl and top with the sauce, **¼ cup [25 g] toasted unsalted sliced almonds** and **½ tsp ground sumac**. Drizzle with **extra-virgin olive oil** and serve hot or at room temperature.

Store leftovers refrigerated in an airtight container for up to 3 days.

One of the most reliable and comforting ways to enjoy any type of starchy vegetable, especially tuberous ones, is to boil and mash it to creamy smoothness. This is a lovely side with braised meats and roasted root vegetables, and a fantastic alternative to mashed potatoes. It is also very good inside a crisp dosa accompanied by the tomato sauce.

THE COOK'S NOTES

- If you can't find yams, use russet potatoes; it will be different but still good.

- I've noticed that mashed yams are like sponges when it comes to flavor and often require a strong hand with seasoning. Double, triple, or quadruple the number of fresh garlic cloves needed to suit your tastes.

Lemon + Artichoke Yams

MAKES 4 SERVINGS

Add **1½ lb [680 g] yams, peeled and cut into ½ in [13 mm] cubes**, and **1 tsp fine sea salt** to a large saucepan filled with enough water to cover the yams by 1 in [2.5 cm] and bring to a boil over medium-high heat. Lower the heat to a simmer, cover, and cook until the yams are tender and easily pierced by a fork or knife, 12 to 15 minutes. Remove from the heat and drain into a colander over the sink. Keep the yams in the colander until ready to use.

While the yams cook, prepare the artichoke sauce. In a medium saucepan, heat **2 Tbsp extra-virgin olive oil** over medium heat. Add **4 garlic cloves, thinly sliced**; **1 Tbsp capers, drained**; **1 tsp red pepper flakes such as Aleppo, Maras, or Urfa**; and **1 tsp smoked sweet paprika**. Sauté until fragrant, 30 to 45 seconds.

Stir in **one 14 oz [400 g] can artichoke hearts packed in water, drained and chopped**; **zest of 1 large lemon**; and **3 Tbsp fresh lemon juice**. Cook until fragrant, 30 to 45 seconds. Stir in **½ cup [120 ml] water** and **1 cup [60 g] grated Parmesan**.

Fold in the hot cooked yams. Season with **fine sea salt** and **pepper**, and **additional lemon juice as desired**. Garnish with **2 Tbsp chopped chives** and drizzle with **extra-virgin olive oil**, if desired. Serve immediately.

Many good recipes venture far from their original conceptions. This started out as a pasta recipe, but while researching and learning how to cook yams, I realized that this tuber's somewhat neutral flavor makes it a perfect substitution for a variety of dishes. Boiled yams draw in citrus flavors and spices like a magnet, while the artichokes provide a silky smooth and tender textural counterpart to the starchy yam flesh.

THE COOK'S NOTES

- If you can't find yams and want to make this recipe, you can sub in russet potatoes, carrots, or parsnips.

- If you want this a little spicier, use a hotter variety of red pepper, such as cayenne or Calabrian.

- For a more pronounced caper flavor, double the quantity of capers, drain, and fry half of them in hot olive oil until crisp and golden brown. Fold them in at the final stage.

Sweet + Sour Yams

MAKES 4 SERVINGS

Preheat the oven to 350°F [180°C]. Line a baking sheet with foil.

In a large bowl, toss **1½ lb [680 g] yams, peeled and cut into ½ in [13 mm] cubes**, with **2 Tbsp extra-virgin olive oil** and **fine sea salt**. Spread on the prepared baking sheet. Cover with a second layer of foil and crimp the edges to form a tight seal. Bake for 30 minutes. Flip the yam pieces with a silicone spatula (don't bother with a pair of kitchen tongs; there are too many pieces to turn). Bake, uncovered, for another 30 to 45 minutes, until golden brown and crisp. Remove and transfer to a serving bowl.

While the yams roast, add to a blender or food processor **½ cup [100 g] packed light brown sugar**, **½ cup [120 ml] rice vinegar**, **1 cup [210 g] diced pineapple**, **1 tsp fish sauce or liquid aminos**, **2 Tbsp fresh lime juice**, and **1 tsp red pepper flakes (use a hot variety)**. Pulse at high speed until smooth. Taste and season with **fine sea salt**. Transfer the sauce to a small saucepan and bring to a boil over medium-high heat. Remove from the heat, pour the hot sauce over the hot cooked yams, and fold to coat.

Garnish with **2 scallions, both white and green parts, thinly sliced; 2 Tbsp chopped cilantro**; and **zest of 1 lime**. Serve immediately.

Store leftovers refrigerated in an airtight container for up to 3 days.

I kept working on a way to make sweet, candied yams, but they just didn't taste as right as they do when made with sweet potatoes. The more I tried, the more I felt I tried too hard, and I eventually concluded that the candied dish works beautifully with sweet potatoes because sweet potatoes are naturally sweet. It's a completely different story when fruity, sour flavors come in with a bit of heat and sweetness, as here. Enter vinegar, pineapples, and hot red pepper. This is a good side for cooler weather but also at home at a summer barbecue.

THE COOK'S NOTES

- If you can't find yams, swap in russet potatoes or carrots. Sweet potatoes will also work.

- The yams will become firm when roasted, and once the hot yams meet the sauce, they will suck up almost all the liquid like a sponge.

- Use a hot variety of red pepper flakes here, not the gentler Aleppo, Maras, or Urfa.

- The acids from the vinegar and pineapple give the sauce the much-needed dose of sharpness to counterbalance the sweetness.

- For my friends who like things saucier in life, double the quantity of the sauce. Use the amount listed in the recipe and then serve the extra on the side.

Bamboo
+ Corn

The Grass Family
POACEAE

Origins
BAMBOO IS FROM ASIA, AND CORN IS FROM SOUTHERN MEXICO.

Bamboo
I am known for picking out bamboo shoots (sprouts) from a stir-fry before it's portioned out, because I love them too much to share. Bamboo shoots are eaten in many parts of Asia. In China, they are called zhú sǔn, and in Japanese takenoko. In India, bastenga, a fermented bamboo shoot preparation, is used to add flavor to various dishes from salads to stews. Edible bamboo is sold in fresh, canned, and dried forms. The younger the shoot, the more tender the texture. I purchase canned bamboo shoots from Asian markets and grocery stores (check the Asian food section). You can also find fresh bamboo shoots at farmers' markets.

Corn/Maize
If there's one vegetable that epitomizes summer, it's sweet corn, harvested when the kernels are at peak ripeness and full of sugar. But that also means that using your corn at its peak is a race against time. Sweet corn must be consumed soon after being picked because enzymes present in the kernels convert the sweet-tasting sugars into tasteless starch. Refrigerating sweet corn helps reduce some of this loss in sweetness.

The sweetest variety is the yellow-colored corn. While the kernels tend to be the focus of many recipes, the husk is extremely valuable. Toast the husk and steep it to make a tea, or use it to make a very "corny" flavored stock (see Sweet Corn Pulao, page 79).

There are other forms of edible corn, like baby and candle corn, which are harvested from immature corn ears. Eat them raw in salads or cook them first. Popcorn does not come from sweet corn; it comes from another variety, called *flint*, that has a harder kernel. Polenta and hominy are also made from flint corn.

Storage
Fresh and cooked bamboo shoots can be stored submerged in water and left in an airtight container in the refrigerator for up to a week. I am not a fan of freezing them, as they lose their crunchy texture.

Husked corn cobs and separated kernels in either raw or cooked form can be frozen and stored in the freezer for up to 3 months. One way to hold on to the flavor of the corn is to blanch them in a pot of boiling water for 5 minutes, then shock the kernels in cold ice water to cool them off and stop the cooking process. Drain and pat dry before refrigerating or freezing. Fresh and cooked corn can be stored for up to a week in the refrigerator. Corn husks must be patted dry and stored in airtight bags or containers before going into the freezer.

Cooking Tips

- Raw bamboo must be cooked before it can be eaten. (Canned bamboo shoots have been cooked; they are ready to eat.) Uncooked raw bamboo tastes extremely bitter due to the presence of toxic substances called *cyanogenic glycosides* (also found in apple pips and cassava). Boiling peeled raw bamboo shoots in salted water for 25 to 30 minutes destroys the toxic substances and tempers the bitterness, leaving the shoots with a pleasantly mild, corn-like flavor. Discard the cooking water, rinse the cooked bamboo shoots well, store them in ice-cold water, and use as needed.

- Rehydrate dried bamboo shoots in a bowl of boiling water; leave to soften for several hours. These aren't as flavorful as fresh or canned versions.

- When making creamy bisques or puréed corn soups (see Creamy Corn Soup with Jalapeño Oil, page 77), first boil the corn kernels in water with salt and baking soda. For 2 ears of corn, use 2 cups [480 ml] water, ½ tsp fine sea salt, and ¼ tsp baking soda. The combined action of heat, water, and the two sodium salts helps make the pectin inside the corn kernels water-soluble. The alkaline condition created by the baking soda also provides two extra benefits: It prevents the starch present inside the kernels from gelatinizing too much— so the soup won't be excessively viscous—and the baking soda acts as a catalyst, promoting development of flavors via caramelization and the Maillard reaction.

- Save those corn husks to make (of course) tamales and also a pot of toasted corn husk stock. Use the stock as you would any stock to make soups (Creamy Corn Soup with Jalapeño Oil, page 77) and flavor your dishes (Sweet Corn Pulao, page 79).

Toasted Corn Husk Stock

This is the simplest version of the Toasted Corn Husk Stock that uses only the husks.

MAKES 4 CUPS [945 ML]

You can build on this basic recipe, adding more flavor boosters like onions, dried mushrooms, bay leaves, cloves, and so on. If you have corncobs to spare, roast them for a few minutes until the kernels slightly char and add them to the water with the husks. Avoid toasting the husks too dark or they will make the stock taste bitter.

Toast the **husks from 2 ears of corn** at 350°F [180°C] for 10 to 12 minutes, until they turn light golden brown. Immediately submerge them in a stockpot with **6 cups [1.4 L] cold water**. Add **1 Tbsp whole black peppercorns** and bring to a boil over high heat, lower the heat to a simmer, and cook until reduced to 4 cups [945 ml] of liquid. Drain in a colander, reserving the liquid. Squeeze the liquid from the husks and discard. Refrigerate for up to 1 week, using as needed. You can freeze the stock and store it for up to 1 month.

Corn, Cabbage + Shrimp Salad

MAKES 4 SERVINGS

Brush **4 ears sweet corn, husked,** with **2 Tbsp neutral oil with a high smoke point such as grapeseed**. Sear the corn in a skillet or using a stove-top griddle until it develops deep char marks all over, turning them over with a pair of tongs, for 10 to 12 minutes total. Transfer the cobs from the skillet to a rack and let rest for 5 minutes, until cool enough to handle. Strip and collect the corn kernels from the cob by slicing with a knife. Transfer the kernels to a large bowl.

In a large saucepan or stockpot over medium-high heat, bring to a rolling boil **4 cups [945 ml] water** and **1 tsp fine sea salt**. Prepare a medium bowl of ice water.

To the boiling water, add **1 lb [455 g] medium shrimp, peeled and deveined**.

Cook the shrimp in the salted water until they turn pink, about 3 minutes. Remove the shrimp with a slotted spoon and transfer them to the ice water bath until they become cold, about 5 minutes. Remove the shrimp with a slotted spoon and add to the corn, along with **1 cup (3½ oz [100 g]) finely shredded green or red cabbage**; **½ cup [90 g] cooked freekeh or pearl barley, cooled**; **1 shallot, cut in half lengthwise and thinly sliced**; **2 scallions, both white and green parts, thinly sliced**; **1 garlic clove, minced**; **1 fresh chilli such as jalapeño or serrano, thinly sliced**; **¼ cup [5 g] tightly packed cilantro leaves**; **2 Tbsp chopped fresh mint leaves**; **½ tsp ground black pepper**; and **zest of 1 lime**.

In a small bowl, whisk until combined **3 Tbsp toasted sesame oil, 3 Tbsp fresh lime juice**, and **2 Tbsp honey or maple syrup**. Pour this over the salad. Toss to coat well. Taste and season with **fine sea salt** and **finely ground black pepper** if needed. Serve immediately. Leftovers can be stored in the refrigerator for up to 2 days.

When I was a kid in India, our family ate a lot of seafood, so I quickly learned that shrimp, corn, and cabbage are meant to be in a relationship. Cold shrimp, corn, and cabbage salad was a dish I looked forward to every summer. This is my take on that beloved salad with more fresh herbs, toasted sesame, and lip-tingling fresh lime.

THE COOK'S NOTES

- Red cabbage's pink pigment will bleed a little on standing and stain the shrimp. It doesn't affect the taste, just turns the dish a rosy pink hue.

- Four ears of sweet corn, each approximately 8 oz [230 g], will give you about 3½ cups [525 g] of corn kernels.

A Grilled Corn Feast

Heat a grill to medium-high and oil the grates with a tightly folded paper towel dipped in neutral oil with a high smoke point such as grapeseed.

Brush **4 ears of sweet corn, husked**, with **2 Tbsp neutral oil with a high smoke point such as grapeseed**. Sear the corncobs on the hot grill until they develop deep char marks all over, turning them with a pair of tongs, 10 to 12 minutes. Transfer the cobs to a plate. Use a pastry brush or a butter knife to season the hot corn with any of the following seasonings.

continued

I love sweet corn, and one of my ultimate goals is to host a summertime all-you-can-eat grilled sweet corn party, accompanied by a table offering multiple options of flavored butters, spice blends, and sauces. Until then, I've given you four options that have you covered (see page 70). The Cilantro-Garlic Butter is inspired by Mexican flavors; the Indian Street Food–Style Grilled Corn is packed with flavor from lime, chilli, and salt; the Miso-Garlic Rub is Japanese inspired; and the Sweet Fennel Butter evokes the wild fennel that grows across California. Pick one or pick all.

THE COOK'S NOTES

- If you don't own an outdoor grill or if the summer sun is doing a good job of cooking you, use a grill pan on a stove indoors. It's an excellent backup tool.

- You can use pre-toasted cumin if that's what you have on hand. I don't find it necessary because the heat from the corncob does a wonderful job of releasing the aroma as soon as it meets the butter. If you want to toast your own cumin, carefully toast ½ tsp cumin seeds in a dry cast-iron or stainless-steel skillet over medium heat. The spices will start to turn light brown and release their fragrance in 30 to 45 seconds. Transfer the spices to a small plate and let cool to room temperature before grinding.

- If you prefer a stronger flavor for the Miso-Garlic Rub, double the garlic.

- Each of these seasonings can be made in advance and is enough for four cobs of corn. The flavored butters and miso rub can be frozen in airtight containers or wrapped tightly with cling film.

- Leftover butter can be stored in an airtight container in the refrigerator for up to 3 days. Extra dry spice rub can be stored in an airtight container in a cool dark spot of the pantry for up to 3 months.

Cilantro-Garlic Butter

MAKES A LITTLE UNDER ½ CUP [100 G]

In a small bowl, combine **¼ cup [60 g] unsalted butter**, at room temperature; **1 fresh chilli such as jalapeño or serrano, minced**; **2 Tbsp chopped cilantro**; **2 garlic cloves, grated**; **½ tsp ground cumin** (see the Cook's Notes); and **flaky salt** as needed.

Indian Street Food–Style Grilled Corn

MAKES ¼ OZ [7 G] SPICE RUB

In a small bowl, mix **1 tsp Kashmiri chilli powder (or ¾ tsp smoked paprika + ¼ tsp ground cayenne)**, and **½ tsp fine sea salt**. Dip **2 limes, cut in half**, in the spice mix. Use the lime as a brush to rub the seasoning onto the corncobs, then squeeze the juice over each.

Miso-Garlic Rub

MAKES ABOUT ⅓ CUP [100 G]

In a small mixing bowl, combine **2 Tbsp extra-virgin olive oil**; **2 Tbsp white miso**; **1 garlic clove, grated** (see the Cook's Notes); and **½ tsp ground black pepper**. Stir in **2 Tbsp boiling water** and mix until smooth.

Sweet Fennel Butter

MAKES ABOUT ¼ CUP [60 G]

Heat a small dry cast-iron or stainless-steel skillet over medium heat. Add **1 tsp fennel seeds** and **10 black peppercorns**. Toast for 30 to 45 seconds, until fragrant. Remove from the heat and transfer to a small plate or bowl to cool. Once cool, grind the mixture to a powder using a mortar and pestle or a spice mill. Add the ground spices to a small bowl along with **¼ cup [60 g] unsalted butter, at room temperature**; **1 Tbsp runny honey or maple syrup**; and **flaky salt** as needed. Fold to combine.

Bamboo Shoot Sesame Salad

MAKES 4 SERVINGS

Peel and trim the ends of **2 large English cucumbers**. Cut them in half lengthwise, then scoop out and discard the seeds. Cut crosswise into ¼ in [6 mm] slices. Add the cucumber to a large mixing bowl. Sprinkle with **1 tsp fine sea salt**. Toss and let stand for 30 minutes. Drain the liquid, rinse the cucumbers with running tap water, drain well, and pat dry with a clean kitchen towel. Wipe the bowl down and return the cucumbers to the bowl.

Add **1 large red bell pepper, cored and cut lengthwise into ¼ in [6 mm] strips**; **one 7 oz [200 g] can sliced bamboo shoots, drained and rinsed**; and **20 fresh mint leaves, torn**.

Prepare the dressing. Heat **¼ cup [60 ml] neutral oil with a high smoke point such as grapeseed** in a small saucepan over medium heat. When the oil is hot, add **1 Tbsp sesame seeds** and **1 tsp red pepper flakes such as Aleppo, Maras, or Urfa**. Swirl until the seeds sizzle and the oil turns orange and fragrant, 30 to 45 seconds. Pour the mixture into a small heat-safe mixing bowl.

Fold into the dressing **½ cup [70 g] chopped roasted unsalted cashews**; **2 Tbsp fresh lime juice**; **1 garlic clove, grated**; **1 tsp anchovy paste** (see the Cook's Notes for the vegan version); and **1 tsp dark brown sugar**. Pour the dressing over the vegetables in the bowl. Toss to coat well, taste, and season with **fine sea salt**. Serve immediately.

This is a refreshing salad that goes well with the Asparagus, Shrimp + Pancetta Fried Rice (page 89). If you don't mind cucumber in more than one dish at a meal, serve it with the Green Beans + Cucumber Noodles (page 237).

THE COOK'S NOTES

- The cucumber is salted for 30 minutes, then well drained, rinsed, and drained again. The salt draws out excess water through osmosis.

- Prepare this salad just before serving. If left too long, the fresh mint will blacken and go limp. It won't affect flavor, just aesthetics.

- There isn't a perfect vegan substitute for anchovy paste in this recipe. However, this option gets pretty close: mix **1 tsp white or yellow miso paste** with **1 tsp liquid aminos** and add that to the dressing.

Kimchi Creamed Corn

MAKES 4 SERVINGS

In a large cast-iron or stainless-steel skillet over medium heat, warm **2 Tbsp extra-virgin olive oil**. Add **1 large yellow or white onion, diced**, and sauté until it becomes translucent and just starts to brown, 3 to 4 minutes. Add **4 garlic cloves, minced,** and **½ tsp ground turmeric**. Cook until fragrant, 30 to 45 seconds.

Stir in **3½ cups [560 g] corn kernels** and sauté for 3 to 4 minutes, until the corn is tender; a kernel should taste sweet and soft without a starchy texture. Stir in **1 cup [240 ml] plain, unsweetened full-fat coconut milk** and simmer until the liquid reduces by half, 2 to 3 minutes.

Fold in **2 packed cups (7 oz [200 g]) chopped cabbage kimchi** and **1 Tbsp apple cider vinegar** and cook until the kimchi is warmed through.

Remove from the heat, transfer to a serving bowl, and garnish with **2 Tbsp chopped chives or cilantro**. Serve hot or warm. This dish can be prepared a day ahead of time and refrigerated in an airtight container. Warm before serving.

This is creamed corn on a flavor fast train with a final destination of Absolute Joy. The sweet corn kernels swim in a rich concoction of coconut milk and kimchi, making it equal parts sweet, savory, salty, and bright. This is a dish you might want to serve with bread on the side to mop up every drop of flavorful sauce. It's also vegan—and if you invite me over for a Thanksgiving potluck, this is the dish I'll bring along.

THE COOK'S NOTES

- My favorite brand of coconut milk is Aroy-d, and I can't thank enough Chef Cheetie Kumar of Garland restaurant in North Carolina, who first introduced me to this spectacular brand. The flavor is flawless and as close as you can get to fresh milk straight from the coconut.

- Turmeric might seem like an unusual addition to the kimchi, but here it adds a bright tone to the milk.

Braised Bamboo + Mushrooms

MAKES 4 SERVINGS

In a large saucepan or wok over medium-high heat, warm **2 Tbsp sesame oil**. When the oil is hot, add **1 lb [455 g] whole cremini mushrooms**; **one 5 oz [140 g] can sliced bamboo shoots, drained and rinsed**; and **1 shallot, thinly sliced**. Cover with a lid and cook until the mushrooms are tender, 5 to 6 minutes, stirring occasionally.

In a small bowl, mix together **2 Tbsp low-sodium soy sauce**; **1 garlic clove, grated**; **1 tsp light or dark brown sugar**; and **½ tsp ground black pepper**. Pour over the vegetables and stir to coat well. Cook until the liquid is absorbed. Taste and season with **fine sea salt**.

Leftovers can be stored in an airtight container in the refrigerator for up to 3 days.

Cremini mushrooms are white mushrooms and baby bellas are portobellos; they're just picked at a different stage of maturity. When braised, they become succulent and juicy and provide a softer counterpart to crunchy slices of bamboo shoots. This is excellent over plain rice or as a side to the Green Beans + Cucumber Noodles (page 237).

THE COOK'S NOTES

- If you want a stronger sesame flavor, don't swap in toasted sesame oil when heating; its smoke point is much lower than that of regular sesame oil (regular sesame oil is 410°F to 446°F [210°C to 230°C] while toasted sesame oil 350°F to 410°F [180°C to 210°C]). The toasted oil will taste bitter. Instead, drizzle a few drops of toasted sesame oil on the dish when it's ready to serve.

- You can swap in oyster or shiitake mushrooms instead. If the mushrooms are large, chop them into bite-size pieces.

Creamy Corn Soup with Jalapeño Oil

MAKES 4 SERVINGS

Jalapeño Oil

Place **1 jalapeño, chopped** in a small heatproof jar or bowl.

In a small saucepan, warm **½ cup [120 ml] neutral oil with a high smoke point such as grapeseed** to 200°F [95°C]. Pour the warm oil over the jalapeño in the jar, stir, and let sit, covered, for at least 60 minutes and up to 4 hours before using. Strain and discard the jalapeño solids before using the oil. This oil can be made a week in advance. Store refrigerated in an airtight container.

To prepare the soup, strip the kernels from **2 ears sweet yellow corn, husks reserved**, and add to a medium saucepan. Using the husks, prepare the corn husk stock (Toasted Corn Husk Stock, page 65).

Add to the kernels **2 cups [480 ml] water**, **½ tsp fine sea salt**, and **¼ tsp baking soda**. Bring the mixture to a boil over medium-high heat. Turn down the heat to low. Add **10 black peppercorns**. Cover and simmer for 30 minutes. The kernels will turn bright yellow.

Transfer the kernels and liquid from the saucepan to a blender. Add **2 cups [480 ml] corn husk stock** and **2 garlic cloves**. Blend until smooth.

Wipe the saucepan clean and place over medium-low heat. Warm **2 Tbsp neutral oil with a high smoke point such as grapeseed**. Add **½ tsp ground coriander** and **20 strands of saffron**. Sauté until the spices are fragrant, 30 to 45 seconds. Strain the soup through a fine mesh strainer held directly over the saucepan. Discard any solids left behind in the strainer. Stir and bring to a boil over medium heat. The soup should be thick and creamy; adjust the consistency with more corn husk stock as needed. Stir in **1 Tbsp lime juice**. Taste and season with **fine sea salt**.

To serve, divide the soup among four serving bowls. Garnish each bowl with **2 tsp crème fraîche or sour cream** and **1 tsp chopped chives**. Drizzle with **1 to 2 tsp jalapeño oil**. Serve hot or warm.

Leftover soup can be stored in an airtight container in the refrigerator for up to 3 days.

continued

It's true that recipes are food science experiments, but this recipe felt particularly educational. The multiple lessons I learned while developing this reminded me of the time I skipped one class and felt they'd taught everything I needed to know that one week. Even if science isn't your thing, this soup is dreamy enough that I implore you to cook it. The sweetness of corn comes through this fragrant soup and the jalapeño oil gives a delightful burst of concentrated green, herby chilli flavor. A few crushed tortilla chips make a crunchy topping.

THE COOK'S NOTES

- Corn provides sugar for sweetness and starch for achieving the creaminess of the soup.

- The addition of baking soda (1) helps break down the corn's pectin to contribute a smoother finish, (2) helps accelerate caramelization and the Maillard reaction, and (3) builds flavor.

- Everyone has their own heat-level tolerance when it comes to chillies. If you prefer more heat, use a serrano instead of a jalapeño. If you prefer milder heat, either use half a chilli or a milder variety, or discard the seeds and midribs, where so much of the hot capsaicin is concentrated.

- Chillies have several fat-soluble parts, including capsaicin, other chilli flavor molecules, and even the green chlorophyll pigment. This means that the warm oil acts as a solvent and draws out those substances from the chilli. The longer the chilli sits in the oil, the stronger the flavor will become. I make my jalapeño oil up to 24 hours in advance, though 1 hour is sufficient for a flavorful oil.

Sweet Corn Pulao

MAKES 4 TO 6 SERVINGS

In a fine mesh sieve, rinse **2 cups [400 g] basmati rice** until the runoff water is clear. Tip the rice into a medium bowl and add enough water to cover by about 1 in [2.5 cm]. Soak for 30 minutes.

Heat a grill to medium-high heat and oil the grates with a tightly folded paper towel dipped in **neutral oil with a high smoke point such as grapeseed**.

Brush **2 ears sweet corn, husked**, with **2 Tbsp neutral oil with a high smoke point such as grapeseed**. Sear the corncobs until they develop deep char marks all over, turning them over with a pair of tongs after 10 to 12 minutes. Transfer the cobs to a rack and let rest for 5 minutes, until cool enough to handle. Strip the corn kernels from the cob by slicing with a knife and collect them in a bowl.

Heat **1 Tbsp neutral oil with a high smoke point such as grapeseed** in a large heavy-bottomed saucepan or Dutch oven over medium heat. Add **2 in [5 cm] fresh ginger, peeled and cut into matchsticks; 4 whole green cardamom pods, lightly cracked; 1 tsp whole cumin seeds; 1 tsp red pepper flakes;** and **½ tsp ground black pepper**. Fry until fragrant, about 1 minute.

Drain the soaked rice and add it to the saucepan. Fry the rice for 2 to 3 minutes, until the grains are coated with the oil and spices and stop sticking to each other.

Add the reserved grilled corn kernels and husks, **4 cups [945 ml] water**, and **2 Tbsp fresh lemon or lime juice**. Season with **fine sea salt**. Bring the water to a rolling boil over high heat, then lower the heat to a simmer, cover, and cook until all the water is absorbed, 10 to 12 minutes. Remove from the heat and let rest, covered, for 5 minutes. Remove and discard the corn husks.

While the rice cooks, in a medium saucepan over medium heat, warm **2 Tbsp neutral oil with a high smoke point such as grapeseed**. Add **4 shallots, peeled, trimmed, and thinly sliced**, and sauté, stirring often, until golden brown, about 12 minutes. If the shallots begin to scorch, deglaze with **1 Tbsp water** and scrape the bottom of the pan. Add **1 large green bell pepper, cored and diced**, and **1 fresh green chilli such as jalapeño, chopped**. Sauté until softened, 8 to 10 minutes. Season with **fine sea salt**.

When ready to serve, transfer the rice to a serving bowl and top with the shallot and bell pepper mixture. Serve warm.

continued

Pulaos or pilafs and pasta have a lot in common: A single dish that goes a long way to feed hungry stomachs is transformed each time depending on what ingredients are used to make it. Pulaos can be eaten on a weekday or included (as a side) in a dinner menu for a celebration. This highly fragrant pulao gets its aroma from a combination of green cardamom, cumin, and corn husks, yet it's the flavors of sweet corn, lemon, and bell peppers that star in this dish. While pulao is a standalone dish, a bowl of plain salted yogurt or your favorite Indian pickle would be most welcome (I like to eat this with lime or tomato achar).

THE COOK'S NOTES

• Unlike the Toasted Corn Husk Stock (page 65), this recipe does not call for toasting the husks. I've noticed that the acid from the lemon heightens the bitterness in the toasted husk, which doesn't taste nice.

Asparagus

The Asparagus Family
ASPARAGACEAE

Origins
ASPARAGUS HAILS FROM THE EASTERN MEDITERRANEAN AND ASIA MINOR.

Asparagus
One of the most versatile vegetables to emerge in spring is asparagus. It must be picked when young or it will become woody. I am not going to ignore the pink elephant in the room: the phenomenon of asparagus pee. Some of you may never have smelled it, as not everyone can detect this distinctive smell that shows up after eating asparagus. Some folks can't pick up the smell due to genetic variations. Some people don't metabolize the chemicals in asparagus to the same degree that others do, and consequently don't smell the smell. It takes 15 to 30 minutes for the "asparagus pee" smell to show up after eating; drinking plenty of water helps, but the smell usually dissipates within 24 to 48 hours.

The yucca plant is a member of this family and is seen in many Southern California gardens and throughout Mexico. It should not be confused with yuca or cassava (see page 210).

Storage
Asparagus is best eaten the day it's picked, so to get the most out of it, cook it the day you buy it. Because asparagus spears are simply stems, treat them as you would freshly cut flowers. Trim the dried bottoms and place them cut side down in a glass or jar containing 1 in [2.5 cm] of water. Cover loosely with a plastic bag, and keep in the refrigerator for 4 to 6 days. Change the water if it becomes cloudy.

Cooking Tips

- Asparagus spears are tender when young. As they age, the thickness of the spear increases, but so does its toughness. The easiest way to get rid of the tough ends is to trim them off with a knife. There is one other method, called the "bend and break" method that I, in all honesty, find impractical and wasteful. Let me explain why. This method depends on knowing where this so-called breaking point exists in the spear, the point where tender transitions to tough. Speaking from experience, depending on how much "gentle" pressure you apply, that spear can break at any random point, and you'll end up with asparagus spears of varying lengths. Looking for this imaginary spot wastes a lot of good asparagus stem, and frankly, trimming off the dry tough end is the most efficient. Another option is to peel the thicker ends before using the asparagus.

- I typically avoid buying and cooking the extremely thin pencil-like asparagus spears. Because they're so thin, they overcook easily. Conversely, the spears that are too thick are often too woody and fibrous and a waste of money.

- Besides becoming tender, green asparagus takes on a brilliant green color when cooked. A pinch of baking soda added to boiling water will help improve the brilliance of the green color, but be careful: If left in the water too long, the asparagus will become very soft. Purple asparagus will lose some of its purple color on heating because the anthocyanins responsible for that color are heat sensitive.

- **Steaming:** Steam asparagus using a steamer insert or a bamboo steamer fitted inside a pot or wok with about 2 in [5 cm] of boiling water. Cooking time will vary depending on the thickness of the asparagus stems. Start checking for tenderness after 5 minutes of steaming.

- **Boiling:** Asparagus spears can be added to a pot of boiling salted water (4 cups [945 ml] water and 1 tsp sea salt) in a large saucepan and cooked until the spears are tender. Check thinner asparagus spears at 2½ to 3 minutes and thicker spears after 5 minutes. Lift out the cooked asparagus spears with a slotted spoon and submerge them in a large bowl filled with ice-cold water to stop them from overcooking. I sometimes add the peel of a lemon (leave the white pith out or it will taste bitter) or a few makrut lime leaves to the boiling water to impart a citrus aroma. Fragrant herbs and spices can also be added to the boiling water.

- **Roasting:** Toss asparagus with a little extra-virgin olive oil or another oil with a high smoke point and salt. Roast the asparagus on a baking sheet or dish at 400°F [200°C] for 25 to 30 minutes, until they turn bright green and begin to char.

- **Grilling:** Just like roasting, grilling is wonderful at developing bitter-sweet and charred flavors. Toss the asparagus with oil and salt, then cook over the hot grates of a grill at medium-high heat until slightly charred and tender, 5 to 6 minutes. Flip them over as needed with a pair of kitchen tongs to allow even cooking. A vegetable grill pan with perforated holes makes the job of grilling easy. If you don't own a grill, consider using a cast-iron grill pan on the stove.

Asparagus, New Potatoes + Sauce Gribiche

MAKES 4 SERVINGS

Add **10 oz [285 g] baby new potatoes, scrubbed and halved lengthwise**, to a medium saucepan. Top with enough water to cover the potatoes by 1½ in [4 cm]. Add **fine sea salt**. Bring the water to a rolling boil over high heat. Lower the heat to a simmer and cook until the potatoes are fork-tender, 15 to 20 minutes. Cooking time will vary depending on the type and size of the potatoes.

Sauce Gribiche

Separate the whites and yolks of **4 large hard-boiled eggs, at room temperature**. In a medium bowl, using a fork, mash the yolks with **1 Tbsp whole-grain Dijon mustard**. Drizzle ⅓ **cup [80 ml] extra-virgin olive oil** into the yolks in a slow, steady stream, whisking with a fork until the mixture emulsifies and becomes creamy. Whisk in **1 Tbsp sherry or white wine vinegar**. Finely chop and add the egg whites. Then add **3 cornichons, minced**; **1 shallot, minced**; **1 garlic clove, minced**; **1 Tbsp brined capers**; **2 Tbsp chopped flat-leaf parsley**; and **1 tsp ground black pepper**. Taste and season with **fine sea salt** and extra **vinegar** if needed.

Drain the potatoes and place them in a large bowl. Gently smash them with a wooden spoon, just enough to break them into bite-size pieces. Season with **a little fine sea salt**.

Set a steamer insert or bamboo steamer lined with a sheet of parchment paper or a few cabbage or lettuce leaves over a large saucepan or wok filled with about 2 in [5 cm] of boiling water over medium-high heat. Add **1 lb [455 g] asparagus, tough ends trimmed and discarded**. Steam until the asparagus turns bright green and tender, 4 to 5 minutes. Remove the asparagus with a pair of kitchen tongs and add them to the potatoes.

Add the sauce gribiche to the asparagus and potatoes. Fold to coat well. Taste and season with **fine sea salt** if needed. Serve warm or at room temperature. Leftovers can be stored for up to 3 days in the refrigerator. I like to add **1 to 2 Tbsp boiling water** to rehydrate the potatoes before reheating and serving.

continued

Sauce gribiche is one of the classic French sauces that is served cold, and here it does a magical number with bright green spears of steamed asparagus and tender baby new potatoes. Expect bites of garlic, salt, fresh herbs, and a good amount of brightness from the vinegar. This is going to come across as sacrilegious to some, but I really love this asparagus dish inside a warm, toasted everything bagel. I think "satisfying" is an appropriate qualifier for my justification. If you eat seafood, serve this with roasted or smoked fish like salmon, mackerel, or trout.

THE COOK'S NOTES

- The Dijon mustard and hard-boiled yolks help emulsify the sauce gribiche and hold it together.

- Cornichon lovers, feel free to play around with the amount listed here to make your hearts happy. And if you do go the bagel route, serve a few extra on the side.

88

Asparagus, Shrimp + Pancetta Fried Rice

MAKES 4 SERVINGS

In a medium bowl, toss together **1 lb [455 g] medium shrimp, peeled and deveined**; **1 Tbsp cornstarch**; **½ tsp fine sea salt**; and **½ tsp ground black pepper**. Set aside.

In a small bowl, whisk together **2 large eggs**, **¼ tsp ground black pepper**, and **⅛ tsp fine sea salt**.

Set a wok over high heat and pour in around the sides **1 Tbsp neutral oil with a high smoke point such as grapeseed**. Lower the heat to medium-high and add the egg to the hot wok. Stir the eggs in a circular motion, scraping the sides of the wok as it cooks, until the egg scrambles. Remove and transfer to a large bowl.

Wipe the wok down and reheat over high heat. Add **4 oz [115 g] diced pancetta**. Stir-fry until the fat renders and the pancetta browns and becomes crisp. Transfer the pancetta to the large bowl with the eggs and leave the fat in the wok.

Add **1 shallot, minced**; **6 scallions, both white and green parts, thinly sliced**; and **2 in [5 cm] fresh ginger, peeled and cut into thin matchsticks**. Stir-fry until the scallions start to brown and the ginger becomes brown and crisp, 3 to 4 minutes. Add **1 lb [455 g] asparagus, tough ends trimmed and discarded, cut into 1 in [2.5 cm] pieces**. Stir-fry until the asparagus starts to turn bright green and char, 3 to 4 minutes. If at any point you think the wok needs a little oil, drizzle in **1 Tbsp grapeseed oil**. Transfer the mixture to the large bowl with the eggs and pancetta.

Wipe down the wok and add **1 Tbsp grapeseed oil**. Heat over high heat. Add the shrimp and stir-fry until the shrimp turns pink, 1 to 2 minutes. Transfer to the large bowl.

Wipe the wok clean and add **1 Tbsp grapeseed oil**. Heat over high heat, then add **5 cups [510 g] day-old, cooked rice**. Break up any lumps and stir-fry for 3 to 5 minutes. Cook until the rice starts to crisp and caramelize a little. Fold in the eggs, pancetta, vegetables, and shrimp.

In a small bowl, whisk together **1 Tbsp low-sodium soy sauce**, **1 Tbsp rice wine vinegar**, and **½ tsp ground black pepper**. Pour the liquid over the rice and vegetables in the wok and fold to coat well. Stir-fry for 1 additional minute. Remove from the heat and serve warm. Leftovers can be stored in the refrigerator in an airtight container for up to 3 days.

continued

Fried rice will always be my favorite dish to order from Chinese restaurants. The beauty of this dish is that it can exist in myriad versions and is one of the most remarkable and inventive ways to use leftovers. Here, the nutty flavor of asparagus is complemented by crispy bits of pancetta and sweet and savory shrimp. Two of my favorite sauces to serve alongside are equal parts soy sauce and rice vinegar mixed with one sliced bird's eye chilli, and a jar of chili crisp.

THE COOK'S NOTES

- While you can get away with using a long-grain rice such as basmati (as is sometimes done in India) or jasmine in place of short-grain rice, what you can't skip is using day-old rice. Freshly cooked rice will be richer in moisture than day-old rice, making it difficult to stir-fry, and it ends up sticking to the wok. In day-old rice, the starch forms crystals that make the grains stronger and ready for stir-frying.

- Pancetta takes the place of lap cheong or the dried smoked pork sausage that is used in classic Chinese cuisine when making fried rice. Make sure to cook the pancetta well so all the fat is rendered and flavors the dish throughout.

- As far as I know, there aren't any good non-pork substitutes for pancetta; however, you can get close to that salty flavor by using diced green olives. If you go this route, you will need to add oil when cooking the shallot; 2 to 3 Tbsp should make up for the missing pancetta fat.

- If, for some reason, you have a leftover omelet in your refrigerator, skip the scrambled eggs in this recipe. Instead, cut the omelet into fine shreds and fold it into the fried rice in the last step.

90

Asparagus Salad with Cashew Green Chutney

MAKES 4 SERVINGS AND 1¼ CUPS [370 G] CHUTNEY

Cashew Green Chutney

Add to a blender **¾ cup [105 g] whole raw unsalted cashews** and **½ cup [120 ml] boiling water**. Soak for 15 minutes. Add **1 bunch [115 g] cilantro, chopped**; **3 scallions, both white and green parts, halved crosswise**; **1 fresh green chilli such as jalapeño or serrano, stemmed**; **1 Tbsp brined green peppercorns, drained**; **zest of 1 lime**; and **2 Tbsp fresh lime juice**. Blend until smooth. Taste and season with **fine sea salt**. Reserve ½ cup [120 g]. The rest of the sauce can be stored for up to 4 days in an airtight container or frozen in a freezer-safe, airtight container for up to 2 weeks.

Heat a grill pan or cast-iron skillet over medium-high heat. There's enough oil on the asparagus that you shouldn't need to brush the pan before heating. In the pan, toss **1½ lb [680 g] asparagus, tough ends trimmed off and discarded**; **1½ Tbsp extra-virgin olive oil**; and **½ tsp fine sea salt**. Cook the asparagus on the hot grill pan, flipping them with a pair of kitchen tongs, until the asparagus turns bright green and develops charred spots or grill marks, 4 to 6 minutes. Transfer the asparagus to a serving plate.

Add **1 large English cucumber, halved lengthwise, seeds scooped out and discarded with a spoon, and then sliced**, and **1 bunch watercress, trimmed (about 7 oz [200 g])**. Season with **fine sea salt**. Dress the vegetables with the cashew green chutney and serve.

As someone with Indian roots who ascribes to the idea that chutneys and achars are meant for more than just dipping or as a sandwich spread, I feel it is imperative to include chutneys in each book I write—but also give you a couple of different ways to use them. Here, the creamy cashew green chutney acts as a salad dressing for the asparagus and cucumber. Fresh watercress packs a pleasant and mild punch of wasabi flavor.

THE COOK'S NOTES

- Soaking the cashews in boiling water helps soften them quickly and makes for a creamier chutney.

- Roasted potatoes and sweet potatoes are also a good addition here.

- Because this salad contains cucumber, I prefer to dress it just before serving to avoid the release of too much liquid.

- Use baby arugula leaves if you can't find watercress.

Orecchiette with Asparagus + Feta

MAKES 4 SERVINGS

Bring a medium saucepan of water with **1 tsp fine sea salt** to a rolling boil over high heat.

Add **½ lb [230 g] orecchiette**. Cook until the pasta is al dente, or per the package instructions. Transfer the pasta with a slotted spoon to a large serving bowl and reserve 1 cup [240 ml] of the cooking water and keep hot.

While the pasta cooks, heat **2 Tbsp extra-virgin olive oil** in a large saucepan or Dutch oven over medium-low heat. Sauté **2 garlic cloves, thinly sliced**, with **½ tsp ground black pepper** and ½ tsp red pepper flakes, if desired, until the spices become fragrant, about 1 minute. Increase the heat to medium-high and add **1 pint [280 g] cherry or grape tomatoes**. Sauté until the tomatoes start to burst, 4 to 5 minutes, crushing the tomatoes with the back of a wooden spoon and scraping the bottom of the pan. Add **1 lb [455 g] asparagus, tough ends trimmed and discarded, cut into 1 in [2.5 cm] pieces**. Cook until they turn bright green and tender, 3 to 4 minutes. Fold in the hot cooked orecchiette.

In a blender, with the lid's center cap removed and the opening draped with a dish towel to let steam escape, pulse **3½ oz [100 g] feta, crumbled**; **1 cup [240 ml] hot reserved cooking water**; and **½ tsp ground turmeric** until combined. Pour over the pasta and fold to coat well. Fold in **1 preserved lemon peel, thinly sliced into strips**; **2 Tbsp chopped dill leaves**; and **⅓ cup [40 g] pine nuts**. Taste and season with **fine sea salt**. Serve hot or warm.

Warm bits of asparagus, bright sautéed cherry tomatoes, and "small ear" orecchiette pasta are bathed in a luxurious turmeric feta sauce. Crunchy pine nuts and fresh dill are all that is needed to finish off this colorful pasta.

THE COOK'S NOTES

- Preserved lemons are made by letting lemons sit in salt and fresh lemon juice for a few months. Consequently, these lemons are very salty and must be rinsed before use. Some brands will also advise you to get rid of the soft pulpy flesh and use only the peels.

- This is a good place to use a citrus zester when cutting strips of lemon peel. Avoid the bitter white pith found under the thin yellow layer of the lemon rind.

- I prefer untoasted pine nuts for the garnish, but you most certainly can toast them.

Beets Chard + Spinach

The Amaranth Family
AMARANTHACEAE

Origins
BEETS COME FROM THE MEDITERRA-NEAN AND THE MIDDLE EAST, CHARD COMES FROM SICILY, AND SPINACH COMES FROM IRAN.

Beets
I was a dramatic child, and the drama hasn't left me as an adult. There were many times I'd forget that I'd eaten beets, only to think I was dying a few hours later due to red pee. Our inability to digest the red pigment in some beet varieties is the source of this color, and the harmless phenomenon is called *beeturia*. Beets come in a variety of colors, from dark reddish-pink to golden yellow and the beautiful candy stripe or Chioggia beets, which I firmly believe should be reserved only for salads, where they stand out.

Chard
Chard, Swiss chard, and silver beets are varieties of beet with edible leaves. The leaves can vary in hue from green to red while the petiole (stem) color can include white, green, yellow, purple, or red. When these are bundled together in all their colorful glory, the chard is labeled as rainbow chard.

Spinach
At home, I playfully refer to spinach as a "shrinking glory" because a large volume of fresh spinach quickly reduces to a small quantity when cooked. Spinach leaves contain more than 90 percent water, and heating releases the liquid from the leaves, shrinking them. This is why recipes often call for a large amount of fresh spinach. Frozen spinach is fresh spinach that's chopped and then either boiled or blanched and then frozen; consider it mostly preshrunk. One 10 oz [280 g] package of frozen spinach is the equivalent of about 1 lb [455 g] of fresh spinach.

Storage
To store beets, separate the greens from the roots. Greens can be refrigerated, wrapped in a damp kitchen towel or paper towels inside a plastic bag or container, for up to a week, or until they begin to turn yellow. For short-term storage, cut the tops 2 in [5 cm] above the root and store them in the crisper drawer of the refrigerator. For a longer period, store the roots in the refrigerator or, if you have one, a damp, cool spot like a root cellar. Or cook and freeze them: Remove the leaves, rinse, and wipe off any dirt. Boil unpeeled beets until tender, about 30 minutes. Transfer to an ice bath to cool completely, remove the skin, slice the beets as you desire, and place in an airtight freezer-safe bag. Store for up to a year. Chard and spinach can be stored short term in the same way as beets: wrapped in a damp kitchen towel or paper towels inside a plastic bag or container. Refrigerate for up to a week.

Cooking Tips

- Beet greens, chard, and spinach can be used interchangeably in most recipes and even in recipes where bitter leafy green vegetables like kale are used. However, the stems of beets and thicker stems of chard need cooking ahead of the leafy parts.

- Beets do not need to be peeled before cooking; just rinse them gently but thoroughly. If you don't want the skin, peel it after cooking; it will slip right off. Avoid scrubbing beets too roughly or their skin will break.

- Red, pink, and yellow beets contain water-soluble pigments that stain everything they touch. To avoid a colorful mess, wear dishwashing gloves or lightly grease your hands, the surface of the cutting board, and the knife with olive oil. The pigments are soluble only in water, not in oil; they will slide off the oiled surfaces with ease.

- There are two ways to roast beets: low and slow at between 250°F and 350°F [120°C and 180°C] or high and fast at 400°F to 425°F [200°C to 220°C]. I prefer the high and fast method, but wrap the beets tightly with foil to prevent them from drying out. The foil also helps steam and soften the beets (see Chilli Beets + Lima Beans with Cucumber Olive Salad, page 111). Once cool enough to handle, the skin will come off with ease.

- I avoid boiling beets because they can lose their flavor and color. However, in some recipes like pickled beets or the Beets, Toasted Barley + Burrata Salad (page 103), boiled beets taste better than roasted. The pickling liquids and spices also contribute to flavoring.

- Tender raw beet greens are wonderful when chopped and used in salads. They can also be sautéed (Beet Greens, Turmeric + Lentil Risotto, page 105) and used like other leafy greens.

- Avoid buying spinach (or, for that matter, any leafy greens) with yellow leaves or slimy spots. The leaves should be firm and bright green. Frozen spinach is great too. I always keep a bag or two on hand in case there isn't time to shop.

- Compared to other leafy greens, chard leaves don't rip as easily and are great for wraps and making stuffed rolls. They can be used in place of cabbage leaves in the Stuffed Cabbage Rolls in Tomato Sauce (page 174).

- Fresh spinach contains a lot of water and oxalic acid. When cooked at temperatures above 140°F [60°C], the acid breaks down and releases more water. Cooked spinach won't feel like it's etching your teeth the way fresh spinach does.

- There are several different ways to wilt fresh spinach. In Italy and some other European countries, fresh spinach leaves are submerged in boiling water until tender and then drained. Fresh spinach leaves can also be wilted by microwaving or sautéing. In either case, squeeze and drain out as much liquid as possible. Frozen spinach can be thawed and left to drain to remove the excess liquid.

99

Baked Eggs with Tadka Greens

MAKES 4 SERVINGS

Preheat the oven to 350°F [180°C].

In a large cast-iron or other oven-safe skillet over medium heat, melt **1 Tbsp ghee or extra-virgin olive oil**. Add **2 large leeks, trimmed and thinly sliced**. Sauté until golden brown, 4 to 5 minutes. Add **2 garlic cloves, thinly sliced**; **1 tsp red pepper flakes such as Aleppo, Maras, or Urfa**; and **½ tsp ground black pepper**. Sauté until fragrant, 30 to 45 seconds. Add **1 large bunch (about 1 lb [455 g]) chard leaves and stems, shredded**. Sauté until the leaves start to wilt and the stems become tender, 5 to 6 minutes.

Remove from the heat and fold in **one 14 oz [400 g] can pinto beans, drained and rinsed**, and **1 Tbsp fresh lemon or lime juice**. Taste and season with **fine sea salt**.

Using a spoon or spatula, make four wells in the skillet. Crack and drop into the wells **4 large eggs**. Place the skillet in the oven and bake until the egg whites become opaque and the yolks are still runny, 5 to 6 minutes. Remove from the oven.

While the eggs cook, prepare the tadka. In a small saucepan over medium-high heat, melt **2 Tbsp ghee or extra-virgin olive oil**. When hot, add **1 tsp whole caraway seeds**, **1 tsp whole black or brown mustard seeds**, and **1 tsp ground coriander**. Swirl gently until the seeds stop sputtering and turn a light golden brown, 30 to 45 seconds. Remove from the heat and add **½ tsp smoked sweet paprika**.

Ladle the hot oil with the spices over the eggs and greens and serve immediately. This is best eaten within an hour of making; leftovers can be stored for up to 1 day in the refrigerator but keep in mind runny eggs won't reheat well.

Tadka, which goes by many other names, is one of the most spectacular flavor-building techniques used in Indian cooking. In the tadka method, whole or ground spices and other aromatic ingredients such as garlic or curry leaves are dropped into a small quantity of hot fat with a high smoke point. The heat and the oil help draw out the aromatic molecules from the spices to create a flavorful concoction, which is then poured, warm, to finish your dish. Serve it with warm bread and tomato chutney (see Okra Preserved Lemon Tempura with Tomato Chutney, page 253).

THE COOK'S NOTES

- With a tadka, the oil is very hot, and spices can easily burn and turn bitter. I recommend either removing the hot oil from the burner and then adding the spices, or cooking them for only 30 to 45 seconds.

- To test whether the oil is hot enough for the tadka, drop one or two mustard seeds into the oil. If the seeds start to sizzle and pop, the oil is hot enough.

Beets, Toasted Barley + Burrata Salad

MAKES 4 SERVINGS

Fill a large saucepan with enough salted water to cover **8 small beets, scrubbed and trimmed** by at least 1 in [2.5 cm]. Bring to a boil over medium-high heat, lower the heat to a simmer, cover, and cook until the beets are tender and a knife or skewer passes easily through the center, 20 to 40 minutes. Add more water as necessary to keep the beets covered. Remove the beets with a slotted spoon and let cool on a plate. When cool enough to handle, peel the beets and cut each into quarters.

Put the cooked and quartered beets in a small saucepan and add **½ cup [120 ml] rice vinegar**; **¼ cup [60 ml] maple syrup**; **1 tsp ground black pepper**; **1 tsp poppy seeds**; **1 tsp red pepper flakes such as Aleppo, Maras, or Urfa**; and **½ tsp fine sea salt**. Bring to a boil, lower the heat, and let simmer, uncovered, stirring occasionally, until the liquid reduces to approximately ¼ cup [60 ml], 8 to 10 minutes. Remove from the heat and let sit until cool enough to handle. The beets can be prepared at least 3 days in advance and stored in an airtight container in the refrigerator, and then brought to room temperature.

Heat a dry stainless-steel skillet over medium heat. When the skillet is hot, add **2 Tbsp pearl barley**. Toast by swirling the grains in the skillet until they start to turn golden brown and fragrant, 2 to 3 minutes. Transfer to a small plate to cool completely. Crush the barley—with a mortar and pestle or a buzz grinder—to form a coarse mixture.

Assemble the salad by placing together on a serving plate **5 oz [140 g] baby arugula** and **8 oz [230 g] Burrata, at room temperature**. Top with the pickled beets and ground toasted pearl barley. Drizzle with **2 Tbsp extra-virgin olive oil** and sprinkle with **flaky salt**. This salad is best eaten within an hour after assembly, as it does not store well.

continued

When my husband, Michael, traveled to China often, he would come back with all sorts of assorted flavors of teas for us to try. A particular standout was toasted barley tea, for its warm and nutty fragrance. I've borrowed that flavor profile here and used toasted barley not only for its aroma but also for its crunchy texture. The beets are pickled quickly in rice vinegar, sweetened with maple syrup, and served with Burrata and fresh peppery arugula leaves. This is a good salad on its own but is also wonderful to share with people as part of a potluck alongside other dishes like the Crispy Sunchokes + Preserved Lemon Gremolata (page 119) or the Lentil Lasagna (page 242).

THE COOK'S NOTES

- By using smaller beets and boiling them first in water and then in the vinegar mixture, you'll find the beets become tender and are pickled quickly.

- Toasting the pearl barley helps produce a panoply of toasty and nutty aromas with a crunchy texture for this salad. It will absorb liquid easily, so add when you're just ready to serve.

- The red color from the beets will stain the cheese, but that's not the end of the world. If you prefer to control the color, serve the beets on the side or assemble just before serving.

Beet Greens, Turmeric + Lentil Risotto

MAKES 4 SERVINGS

In a medium saucepan over medium-high heat, bring to a simmer **5 cups [1.2 L] Master Mushroom Vegetable Stock (page 337)**. Lower the heat to keep the stock warm. You might end up not using all the stock.

Prepare the risotto by heating in a large saucepan over medium heat **2 Tbsp ghee, extra-virgin olive oil, or unsalted butter**. When hot, add **2 garlic cloves, thinly sliced, and ½ tsp ground turmeric**. Sauté until fragrant, 30 to 45 seconds. Stir in **1 cup [200 g] Arborio rice, rinsed and drained**, and **½ cup [100 g] red lentils, rinsed and drained**. Fold to coat with oil, about 1 minute.

Stir in 1 cup [240 ml] of the stock. Continue to stir until the rice absorbs most of the liquid. Continue to add ½ cup [120 ml] of stock at a time, stirring each addition until the liquid is all absorbed. Repeat until the rice and lentils are tender, 20 to 25 minutes. You will need about 4 cups [945 ml] of the warmed stock. Remove the saucepan from the heat and stir in **1 Tbsp fresh lemon juice**.

Prepare the beet greens. In a large cast-iron or stainless-steel skillet over medium-high heat, warm **2 Tbsp ghee, extra-virgin olive oil, or unsalted butter**. Add **2 garlic cloves, grated**; **1 tsp cumin seeds**; **1 tsp ground black pepper**; **1 tsp red pepper flakes such as Aleppo, Maras, or Urfa**; and **½ tsp ground turmeric**. Sauté until fragrant, 30 to 45 seconds.

Fold in, a handful at a time, **1 lb [455 g] beet greens and tender stems, chopped (from 2 to 3 bunches of beets)**, until the leaves completely wilt and the stems become tender. Add 1 to 2 Tbsp water if the pan seems too dry and the spices start to scorch. Stir in **2 tsp fresh lemon juice**. Taste and season with **additional lemon juice** as desired and **fine sea salt**.

If the risotto becomes firm on cooling, loosen it by stirring in ½ cup [120 ml] or more of the warmed stock as needed. To serve, top the risotto with the cooked beet greens. Garnish with **¼ cup [15 g] grated Parmesan** and **2 Tbsp toasted pine nuts**.

Leftovers can be stored in an airtight container for up to 3 days. Add 1 to 2 Tbsp water to help loosen the risotto before reheating.

continued

I find it fascinating when I notice how similar dishes are prepared in different parts of the world. Take Indian khichdi and Italian risotto, for example: Both rely on the use of starch in rice to create a luxuriously creamy texture, and both fall into the comfort food category. This risotto is a representation of both those dishes through the incorporation of ghee, lentils, and turmeric used in khichdi, while using the Arborio rice, stock, and Parmesan that are more common in risotto.

THE COOK'S NOTES

- This risotto is a wonderful way to use up all those beet greens, but you can use other green leaves like chard, collards, and even kale.

- In this risotto, the starch that thickens the liquid comes from both the rice and the lentils.

- My preference here is for ghee or butter because it gets very close to the flavor of khichdi, but olive oil can be used.

- You can toast your own pine nuts by cooking them in a small stainless-steel skillet over medium heat until the seeds turn golden brown. Be careful not to overcook them or they will become bitter.

106

Crispy Salmon with Green Curry Spinach

MAKES 4 SERVINGS

In a large cast-iron or stainless-steel skillet over medium-high heat, warm **2 Tbsp extra-virgin olive oil**. When the oil is hot, add **2 Tbsp drained capers**. Sauté until crispy and lightly golden brown, 1½ to 2 minutes.

Add **1 large red onion, minced**. Sauté until translucent, 4 to 5 minutes. Add **2 Tbsp green curry paste** and **4 garlic cloves, thinly sliced**. Sauté until fragrant, 1½ to 2 minutes. Add, a handful at a time, **1 lb [455 g] fresh baby spinach**. Sauté, tossing, until the leaves are completely wilted and most of the liquid evaporates, 10 to 12 minutes. Stir in **1 cup [120 g] fresh or frozen peas**, **1 cup [240 ml] unsweetened coconut milk**, **½ cup [120 ml] water**, and **½ tsp ground black pepper**. Bring to a boil and let simmer until the peas turn tender, 1½ to 2 minutes.

Stir in **1 Tbsp fresh lime or lemon juice** and **1 tsp low-sodium soy sauce**. Taste and season with **fine sea salt**. Remove from the heat and cover to keep warm.

Use clean paper towels to pat dry **four 6 oz skin-on salmon fillets**. Season on both sides with **fine sea salt** and **ground black pepper**.

In a large cast-iron or stainless-steel skillet over medium heat, melt **2 Tbsp unsalted butter**. Cook until the butter stops crackling and its water evaporates, 2 to 3 minutes. Add **2 Tbsp extra-virgin olive oil**. Place the salmon fillets on the hot pan, skin side down. Cook without moving the fillets (if moved before the skin is fully cooked, the flesh can easily tear away from the skin). After the first 5 minutes of cooking, tilt the skillet toward you and use a large spoon to collect the oil in the pan and pour it over the fish fillets. Cook until the skin starts to turn golden brown and crispy on the sides, another 1 to 2 minutes. Transfer the cooked fish from the pan to a plate. Serve with the spinach and garnish with a few **sprigs of cilantro or microgreens like arugula or kale**.

continued

My aunt Elaine is a firm believer that salmon should not be used in curries because it becomes unpleasantly firm, unlike many other fish. But what if the salmon is cooked separately and served on the side? This recipe might be tailored for my aunt, but you will enjoy it too. Coconut rice (see Cashew + Bell Pepper Chicken with Coconut Rice, page 295) or plain rice are both excellent options to serve alongside.

THE COOK'S NOTES

- The milk proteins in the butter will help the fish skin cook well and also release easily. The proteins bind to the metal surface of the pan, creating a nonstick surface that will help the fish slide off with ease after cooking.

- If you don't eat fish, make a batch of the Bok Choy and Crispy Tofu (page 166) and serve it on top of the curry. Seared slices of tempeh or fried eggplant are also wonderful options here.

- There are several terrific brands of Thai curry paste available online and in grocery stores. One of my favorites is the Mekhala brand of curry pastes (they're vegan!).

- I sometimes add chunks of squash like pumpkin and other cucurbits like chayote to this curry.

Chilli Beets + Lima Beans with Cucumber Olive Salad

MAKES 4 SERVINGS

Preheat the oven to 425°F [220°C].

Rub **4 small red beets, peeled and quartered**, with **1 Tbsp neutral oil with a high smoke point such as grapeseed**. Arrange them on a baking sheet or roasting pan, cover with a layer of foil, and seal the edges tightly. Cook until tender and a knife or skewer passes through the center with ease, about 30 minutes.

While the beets cook, prepare the salad. In a medium bowl, combine **1 cup [240 g] plain, unsweetened full-fat Greek yogurt**; **1 garlic clove, grated**; **1 Tbsp fresh lemon juice**; and **1 tsp ground black pepper**. Taste and season with **fine sea salt**. Fold in **1 large English cucumber, peeled and sliced ¼ in [6 mm] thick**, and **1 cup [140 g] pitted green olives, drained and halved**. Transfer to a serving bowl and toss gently with **2 Tbsp extra-virgin olive oil** and **1 tsp red pepper flakes such as Aleppo, Maras, or Urfa**. Refrigerate until ready to use.

In a large cast-iron or stainless-steel skillet over medium heat, warm **2 Tbsp extra-virgin olive oil**. When the oil is hot, add **1 garlic clove, crushed**. Sauté until fragrant, 30 to 45 seconds. Add **4 scallions, both white and green parts, thinly sliced**. Sauté until they just start to turn light brown, 3 to 4 minutes. Add **two 14 oz [400 g] cans lima beans or other white beans, such as cannellini, rinsed and drained**; **2 Tbsp chili crisp or hot sauce**; and **¼ cup [60 ml] water**. Cook until the liquid starts to bubble and the beans are warmed through. Remove from the heat and add the roasted beets. Add the **zest of 1 lemon** and **2 Tbsp fresh lemon juice**. Taste and season with **fine sea salt**.

Serve warm alongside the cucumber olive salad. This salad is best eaten within an hour of preparation, but the lima beans and beets can be stored in an airtight container in the refrigerator for up to 3 days.

I've got a strong bias for dishes that play with tempera-ture. This salad is composed of two parts: one served warm and the other cool. Warm beets and lima beans are cooked and served alongside a chilled yogurt salad made with olives and cucumbers. The heat heightens the sweet and spicy flavors while the cooler component softens the garlic and citrus flavors.

THE COOK'S NOTES

- Smaller beets will make the cooking easier and shorter.

- For a fine grate of the lemon peel, use a Microplane zester.

Artichokes Sunchokes Endive Escarole Radicchio + Lettuce

The Sunflower Family
ASTERACEAE

This is the largest family of flowering plants in the plant kingdom.

Origins
ARTICHOKES, ESCAROLE, AND LETTUCE COME FROM THE MEDITERRANEAN. SUNCHOKES ORIGINATED IN NORTH AMERICA. ENDIVES COME FROM ASIA, AND RADICCHIO HAILS FROM ITALY.

Artichokes or Globe Artichokes
If you get the opportunity to drive California's breathtaking Highway 1, I highly recommend taking a trip to Castroville just for a glimpse of their magnificent artichoke fields. Some of them line up along the coast of the Pacific Ocean, making for some truly stunning landscapes. When they are in season, you might even spot some giant purple thistles that were left to bloom. The edible portion of the artichoke is the unopened flower bud; when cooked, the base of the leaves (the heart) and the stem become delicately tender and want only to be devoured.

Sunchokes, Jerusalem Artichokes, or Wild Sunflower
Sunchokes are a type of root vegetable that taste mildly sweet and nutty. I love them roasted (see Crispy Sunchokes + Preserved Lemon Gremolata, page 119) or fried because they get irresistibly crunchy. Sunchokes are sometimes called "fartichokes" because they contain large quantities of a dietary fiber called *inulin* that we humans can't digest. When sunchokes are eaten raw or improperly cooked, naturally present bacteria in our gut come into contact with the inulin, break it down, and produce gas.

Endive, Escarole, and Radicchio
I grouped these three together because they are all bitter to taste, making them excellent additions to salads. All three can be eaten raw, grilled, roasted, or braised. Frisée, a curly variety of endive with pale green to light yellow leaves, is quite elegant in salads. Radicchio's leaves are arguably the most beautiful, with bright purple and white streaks that make a brilliant statement in any salad (see Mixed Bitter Greens Salad, page 129). I love escarole because it's so sturdy yet succulent.

Lettuce
One consequence of popularity is the inevitable onset of monotony. Lettuce seems to have succumbed to this. Lettuce is tossed into salads as a filler or shredded to be used as a garnish, but I firmly believe lettuce can shine on its own without needing any side acts. Lettuce doesn't have to be eaten raw—it can be pickled, just like gherkins and cucumbers, in a solution of vinegar and sugar; grilled for a Caesar salad; or blended into a soup. Lettuce is popular, and that popularity results in a vast number of varieties. Common varieties include coral, butterhead, iceberg, little gem, oak leaf, romaine, loose-leaf, speckled, and celtuce or stem lettuce. I most often use romaine, iceberg, and celtuce for grilling and other heat applications because they don't wilt as easily as other varieties. Celtuce is also great in stir-fries.

Storage
To store artichokes, spray the artichoke heads with a bit of water, place them in a ziptop bag, and refrigerate for up to a week. Sunchokes can be stored in a cool, dark spot of the kitchen; for long-term storage, wrap them in dry paper towels to absorb excess moisture and place them in a ziptop bag and refrigerate for 1 to 2 weeks. To store endives, escarole, radicchio, and lettuce, wrap the leaves in a paper towel, insert it into a ziptop bag, and refrigerate. Replace the paper towel as needed if it feels damp.

Cooking Tips

- Artichokes contain a chemical called *cynarin* that makes everything taste sweet, including water. Cynarin binds to the sweet taste receptors in our mouth, and once we drink water, the cynarin is washed away, freeing the receptors to trigger a sweet taste response. Many wine experts recommend avoiding wine altogether when eating artichokes because of this phenomenon.

- Once cut, artichokes and sunchokes discolor quickly due to the browning enzymes called *polyphenol oxidases*. If you don't plan on using cut vegetables immediately, submerge them in a bowl of ice-cold acidic water (use a good squirt of lemon or lime juice).

- When picking sunchokes, try to select ones that aren't covered in too many knobs, or else peeling becomes wasteful. Scrub well to remove dirt and roots.

- To reduce gassiness with sunchokes, borrow a tip from dosa making: Ground fennel seeds are mixed with dosa batters to help reduce gassiness. I find this method also works

114

when making sunchoke soup. Inulin breaks down between 275°F and 375°F [135°C and 190°C], so cooking at temperatures in this range or above will help reduce gassiness. If all else fails, take Beano or Gas-X.

- Endive, escarole, radicchio, and lettuce leaves can be steamed, grilled, or sautéed. For grilling, brush the leaves lightly with oil; grill them until the leaves develop delightful char marks.

- Always get rid of excess water when working with lettuce and other fresh greens. Use a salad spinner, or lay the washed leaves out on a dry kitchen towel and gently pat them dry before using. Too much water can dilute the impact of vinaigrettes and other dressings.

- I prefer to dress the three bitter leafy greens—escarole, endive, and radicchio—with extra vinegar or citrus juice to mask some of their bitterness. In salads, these vegetables benefit from the addition of crunchy fresh fruit and a generous dose of fresh herbs.

How to Prepare Artichokes

Cleaning artichokes is a bit of a chore and not something I would recommend doing on a busy day. I keep a stash of frozen and canned artichokes that serves me well when I need to add them to pasta or a casserole. However, for all other occasions except steaming (see Steamed Artichokes with Cashew Red Pepper Dip, page 123), here is the method I use. Larger and medium artichokes are a better investment when cleaning; you end up with a thicker stalk, meatier leaves,

and a bigger heart to eat. Be careful with artichokes; the pointy tips of the outer leaves can easily prick your skin just like a cactus. As you proceed with cleaning and processing, rinse them with acidulated water to prevent browning.

1. Fill a large bowl with 8 cups [2 L] cool tap water and add 2 Tbsp lemon juice (lime juice or vinegar will also work). Rub your hands and knife with a little lemon juice; do this often as you prep the artichoke.

2. To prepare each artichoke, pull off and discard the green and brown tough outer leaves, working until you reach the pale yellow core of inner leaves. These are tender. You will notice an indentation where the leaves curve and extend upward. Cut 1 in [2.5 cm] above the notch and remove and discard the top of the artichoke. The purple part of the leaves is also inedible. Dip the artichoke in the lemon-water bath. Shake to get rid of the excess water.

3. Using a paring knife, trim the stalk to remove the outer green tough, fibrous part and expose the central white core. Trim the sides around where the stalk meets the rest of the artichoke, then dip in the lemon-water bath and shake off the water.

4. Using a teaspoon or a melon baller, scoop out the central filamentous region called the choke. Again rinse the artichoke in the lemon-water bath, this time keeping it submerged. In recipes that call for cutting the artichoke in half, I find it's easier to scoop out the choke after cutting.

5. Prepare the remaining artichokes in the same way and keep them submerged in the lemon water until ready to cook. If you'll be cooking them in hot oil, quickly dry them just before cooking.

115

Crispy Sunchokes + Preserved Lemon Gremolata

MAKES 4 SERVINGS

Preheat the oven to 425°F [220°C]. Line a baking sheet with foil.

Preserved Lemon Gremolata

To make the gremolata, place **1 bunch [130 g] flat-leaf parsley, leaves and tender stems**; **3 garlic cloves**; **1 preserved lemon peel**; and **2 tsp fresh oregano leaves** on a cutting board and chop all with a knife until you get a finely minced mixture. Transfer to a small bowl. Stir in **2 Tbsp shredded Parmesan, 2 Tbsp extra-virgin olive oil, zest of 1 lemon,** and **1 Tbsp fresh lemon juice**. Taste and season with **fine sea salt**. Cover and let sit for 30 minutes.

In a large bowl, stir together **1 lb [455 g] sunchokes, scrubbed and cut into ¼ in [6 mm] thick slices; 2 Tbsp extra-virgin olive oil; ½ tsp fine sea salt;** and **½ tsp ground black pepper** to coat well. Spread the sunchoke slices in a single layer on the prepared baking sheet. Roast, flipping the slices halfway through cooking, until golden brown and crisp, 25 to 30 minutes. Transfer the roasted sunchokes to a serving bowl, top generously with the gremolata, and serve immediately.

This recipe indicates four servings, but if you're like me, you'll think twice about sharing. Crispy foods served with herb sauces are one of my favorite indulgences. Here, seasoned slices of sunchokes are baked until crisp and then served with a generous helping of gremolata. The addition of preserved lemon to the garlic and fresh herbs in the gremolata make it extra special.

THE COOK'S NOTES

- Parmesan builds the savory profile of the gremolata, but if you don't consume dairy, use 1 tsp of nutritional yeast and 1 tsp of liquid aminos to build this taste. A good brand of plant-based Parmesan will also work well.

- Preserved lemons can be found online or in the condiment or international aisles of grocery stores. Since the lemons are preserved in salt, make sure to wash the peel before use and discard the inner pulp.

119

Lettuce with Avocado Caesar Dressing

MAKES 4 SERVINGS

Preheat the oven to 350°F [180°C].

On a rimmed baking sheet, toss together **½ French baguette (about 5¼ oz [150 g]), cut into 1 in [2.5 cm] pieces (day-old bread is best)**; **3 Tbsp extra-virgin olive oil**; **½ tsp garlic powder**; and **fine sea salt**. Spread in a single layer and bake until golden brown and crisp, 12 to 15 minutes, rotating the baking sheet halfway through baking. Remove from the oven and let cool. You can make the croutons at least 3 days ahead of time, but you must store them in an airtight container or they will soften.

Avocado Caesar Dressing

To a blender, add **1 small ripe avocado, pitted and peeled**; **1 green chilli such as jalapeño or serrano, deseeded for milder heat**; **½ cup [120 ml] buttermilk, kefir, or plain yogurt**; **¼ cup [5 g] tightly packed cilantro, leaves and tender stems**; **2 Tbsp grated Parmesan**; **1 tsp onion powder**; **1 tsp fresh lime juice**; **½ tsp ground cumin**; and **½ tsp ground pepper**. Pulse on high speed until smooth and creamy, taste, and season with fine sea salt. This should taste herby and might remind you of a creamy salsa verde. The dressing is at its best when made fresh and eaten with an hour.

In a large bowl, toss together **4 romaine hearts or 4 baby gem lettuce heads, leaves separated and roughly torn into 1 in [2.5 cm] pieces**; the croutons; half the dressing; and **½ tsp ground black pepper**. Taste and season with additional dressing, **fine sea salt**, or **lime juice** as needed. Garnish with **¼ cup [15 g] grated Parmesan**. Serve immediately with any extra dressing passed at the table.

continued

This isn't your classic Caesar dressing, nor is it another boring lettuce salad. It's a cross between Indian and Mexican flavors and one of my all-time favorite ways to eat lettuce. Slather that lettuce up and eat away. Just in case I still haven't sold you on how much I love this, allow me to share a secret: I even eat this for breakfast with slices of hard-boiled eggs.

THE COOK'S NOTES

- If you aren't a crouton person, swap in the Crispy Spiced Chickpeas (page 225). If you love both, use both.

- Even though this dressing contains acid to prevent the browning of the avocado, there isn't enough to keep the avocado from turning brown over a longer period. For this reason, I recommend making the dressing the same day you'll be serving the salad.

- To veganize this dressing, use 1 tsp white or yellow miso paste and 1 tsp liquid aminos. A plant-based Parmesan that you enjoy will also work here.

- The onion and garlic powders used here provide a richer savory taste and mellow the bite of the alliums.

122

Steamed Artichokes with Cashew Red Pepper Dip

MAKES 4 SERVINGS

Cashew Red Pepper Dip

In a blender or food processor, combine **½ cup [70 g] whole raw cashews and ½ cup [120 ml] boiling water**. Let soak for 30 minutes.

While the cashews soak, roast **1 medium red bell pepper** directly over the burner flame of a gas stove until charred all over, 1 to 2 minutes. Alternatively, place the bell pepper on a foil-lined roasting pan or baking sheet and broil on high, rotating every few minutes, until charred all over. Remove the stalk and core and discard the seeds. Add the bell pepper with the charred bits of skin to the food processor along with **2 Tbsp peeled and grated fresh ginger**. Pulse until creamy and smooth.

Add **2 Tbsp chopped cilantro leaves and tender stems**, **2 Tbsp fresh lime or lemon juice**, **1 Tbsp maple syrup**, and **½ tsp smoked sweet paprika** and blend until smooth. Stir in **1 tsp red pepper flakes such as Aleppo, Maras, or Urfa**. Taste and season with **fine sea salt**. The dressing should taste mildly sweet and sour with a smoky ginger flavor. This dip can be made a day in advance. Store in an airtight container in the refrigerator, and bring to room temperature before serving.

Prepare **2 large or 4 medium artichokes**: Using kitchen scissors, trim the pointed tips off each petal and rub with a **freshly cut lemon half**. Trim away the end of the stalk of one artichoke and rub with half a lemon. Peel the stalk. Lay the artichoke on its side and cut away the top quarter and discard, then rub the cut surface with the lemon. Repeat with the remaining artichokes.

Fill a large, deep saucepan or Dutch oven with at least 2 in [5 cm] of water and squeeze any remaining lemon juice into the water. Set a steamer basket inside and bring the water to a boil over high heat. Lower the heat to a simmer. Set the artichokes cut side down in the basket, cover, and steam until the meaty base and leaves are tender, 30 to 45 minutes, or until a leaf can be easily pulled away and a knife inserted into the bottom slides in easily, 45 minutes to 1 hour. Carefully remove the artichokes with a pair of kitchen tongs or a spider and transfer to a serving plate. Drizzle with **2 Tbsp extra-virgin olive oil** and sprinkle with **fine sea salt**.

continued

Serve the artichokes with the cashew red pepper dip on the side. Pull off the leaves, dip the stem end of the leaf into the dip, and eat the meaty portion. Remember to avoid the choke. Extra dip can be stored in an airtight container in the refrigerator for 2 to 3 days.

As much as this recipe is about steaming artichokes, the cashews used to make this creamy red pepper dip also deserve a special shout-out. The dip is smoky, creamy, and sweet, and is also great on its own when you make a veggie board for movie night (see Platters, Boards + Tricks, page 340). It can be made a day ahead, a boon when you are preparing a number of dishes.

OK, back to the steaming artichokes. Steamed artichokes are a revelation in the power of a simple technique that can make certain ingredients sing with joy. Steaming transforms the artichoke into tender succulent leaves that are perfect for dunking into the red pepper dip. On a side note, this is also a good combination for toast or a sandwich.

THE COOK'S NOTES

- If you can't find red bell peppers, do not be tempted to use green; they are simply unripened peppers picked before they sweeten and turn red, yellow, or orange, and they sometimes turn bitter on roasting. Yellow or orange peppers can sub, if available.

- You can skip roasting the red pepper if you have a jar of roasted bell peppers. Remember to drain and rinse the jarred peppers before using.

- Hydrating the cashews in boiling water helps soften them and produces a creamier, smoother texture when blended.

- I like to leave some of the charred skin on the pepper during blending, as it adds a nice smoky flavor to the dip. The smoked paprika helps build on that flavor.

Mixed Bitter Greens Salad

MAKES 4 SERVINGS

In a medium mixing bowl, whisk together **3 Tbsp red wine vinegar**, **1 Tbsp fresh lemon juice**, **1 Tbsp Dijon mustard**, **1 Tbsp honey**, and **zest of 1 lemon** until smooth. Slowly drizzle in **¼ cup [60 ml] neutral oil with a high smoke point such as grapeseed** until creamy and emulsified. Add **½ tsp ground black pepper** and **fine sea salt**.

   In a large mixing bowl, toss together **1 medium head radicchio, leaves separated**; **1 medium head Belgian endive, leaves separated**; **1 large (7¾ oz [220 g]) Granny Smith apple, cut in half, cored, and thinly sliced lengthwise**; **1 Tbsp roasted salted pumpkin seeds**; and half of the vinaigrette. Taste and add more dressing if needed. Serve immediately with the remaining dressing passed at the table.

My relationship with bitter foods is complicated, but for some reason I make an exception for bitter greens. I suspect it's because they work with acids so elegantly. This mixed green salad uses fresh radicchio and endive, and is tossed in a quick and easy vinaigrette made with vinegar, lemon juice, and mustard. It is great by itself when you're looking for something light but also a good accompaniment to grilled vegetables like eggplant and squash.

THE COOK'S NOTES

- I rarely use extra-virgin olive oils in emulsions such as vinaigrettes because it immediately makes the dressing taste bitter. If you want to use olive oil, use a tip from my previous cookbook, *The Flavor Equation*. Mix ½ cup [120 ml] extra-virgin olive oil with ½ cup [120 ml] boiling water and let the mixture sit until the two liquids separate. Discard the water and use the olive oil and your vinaigrette won't taste bitter.

- You can toast your own pumpkin seeds, but if you want to use store-bought, that's fine. I'm sure Ina Garten would approve this message.

- Thin shavings of Parmesan over this salad are lovely.

- Candied walnuts are a good substitute for pumpkin seeds, but you can also use both.

- In summer, when stone fruits are in season, I'll add slices of ripe nectarines and peaches in place of apples.

Braised Artichokes + Leeks

MAKES 4 SERVINGS

In a 12 in [30.5 cm] large and deep cast-iron or stainless-steel skillet, warm **2 Tbsp extra-virgin olive oil** over medium-high heat. Add the hearts from **2 large or 4 medium artichokes, cut in half lengthwise** (see How to Prepare Artichoke Hearts, page 115), and **4 large leeks, white parts only, trimmed and cut in half lengthwise**, directly on the hot skillet, cut side down, and cook until lightly browned, 3 to 5 minutes. If space is tight, brown the vegetables in batches and then cook them together.

Add **1 cup [240 ml] low-sodium vegetable or chicken stock**; **1 garlic clove, crushed**; **1 tsp low-sodium soy sauce**; **½ tsp ground black pepper**; and a **large pinch (15 to 20 strands) of saffron**. Bring to a boil. Turn down the heat to low and simmer, covered, until the vegetables are tender, 10 to 12 minutes. Add water as needed during cooking, making sure there is always liquid halfway up the sides of the artichokes. Uncover, increase the heat to medium-high, and cook until the liquid reduces to about one-quarter of its original volume.

Remove from the heat. Drizzle with **1 Tbsp fresh lemon juice**. Taste and adjust with **fine sea salt** and additional lemon juice. Garnish with **2 Tbsp chopped flat-leaf parsley**. Serve warm. Leftovers can be stored in an airtight container in the refrigerator for up to 3 days.

Braising is a cooking method in which food is first seared to develop flavors using caramelization and the Maillard reaction and then cooked in a flavorful liquid like a broth. Artichokes and leeks braise beautifully, and here they're first browned lightly in a little olive oil and then cooked in a broth flavored with garlic, soy sauce, pepper, and saffron. How to eat it? Well, obviously, I'm going to say over rice, but it is also a wonderful side at barbecues and dinners (I don't know why, but artichokes and saffron always seem to scream fancy for some reason).

THE COOK'S NOTES

- Soy sauce helps build the savoriness of this dish while the saffron adds fragrance and a bright orange tinge. Be gentle with the saffron, as too much of it can be overpowering.

- If you don't have time to prepare the artichokes, use frozen or canned artichokes (packed in water).

130

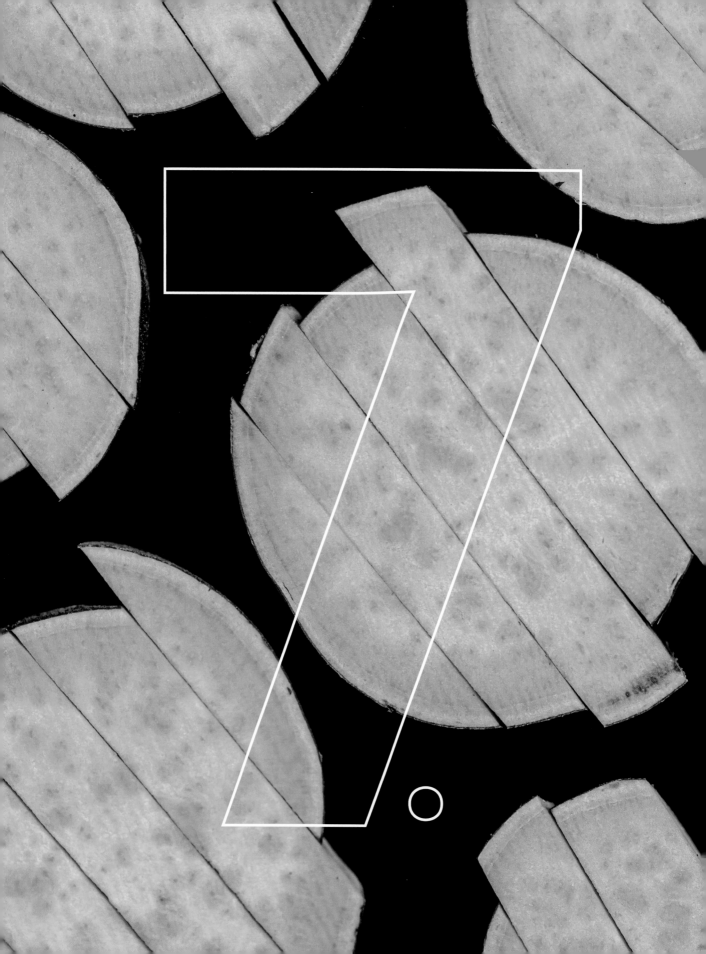

Sweet Potatoes

The Morning Glory Family
CONVOLVULACEAE

Origins
SWEET POTATOES HAIL FROM CENTRAL AND SOUTH AMERICA.

Sweet Potatoes
While sweet potatoes are sometimes referred to as yams, they're not the same thing (see Yams, page 53). They come from different plants; yams are not sweet like sweet potatoes (I find them closer in taste to potatoes), and cooked sweet potatoes have a smoother texture. My general rule of thumb is to avoid swapping these two ingredients in recipes; you're better off using yams in recipes calling for potatoes or yucca. Yams will work well in the Sweet Potato Kale Caesar Salad (page 136) and the Sesame Sweet Potatoes + Gochujang Chicken (page 142).

Storage
Store sweet potatoes in a dark, dry, and cool spot in your kitchen. Sometimes sweet potatoes can start to produce little shoots with leaves. Remove the shoots and cook.

Cooking Tips

- Sweet potatoes come in a variety of colors: orange, white, and purple. The Okinawa or Hawaiian sweet potato with purple flesh is often confused with the purple yam tuber (see Yams, page 53).

- There is a myth that sweet potatoes will burst in the oven if roasted without pricking. This is untrue: The skin of sweet potatoes can breathe and is not a tight waterproof type of skin like that of eggplant. Roast whole, unpricked sweet potatoes in the oven; as they start to cool, the tender flesh will separate from the skin with ease. This holds true for most starchy vegetables that grow in the ground, like yams, potatoes, and sunchokes.

- Avoid boiling or microwaving sweet potatoes; these two cooking methods do not do any justice to this marvelous vegetable. Roasting sweet potatoes in the oven produces at least seventeen different flavor molecules that are not produced with these other techniques.

- When I roast whole or big pieces of sweet potatoes (halves or quarters), I prefer to cook the sweet potatoes twice. First I bake them wrapped in foil. This allows the steam inside the vegetable to make the flesh tender. I then unwrap and roast the sweet potatoes uncovered until the sugars start to bubble and caramelize. This method produces the most wonderful texture and flavor.

- When roasting sweet potatoes that are diced small (see Sweet Potato Kale Caesar Salad, page 136), I skip the first roasting step. Because they're cut so small, there's no benefit from the twice-cooked method, and the sweet potatoes are also done much faster.

- Sweet potatoes are rich in sugar and can burn easily; watch them carefully when cooking at high temperatures.

- The leaves of sweet potatoes are edible and can be sautéed and used in various dishes. Use them in place of beet greens or add some to the beet greens when making the Beet Greens, Turmeric + Lentil Risotto (page 105).

Sweet Potato Kale Caesar Salad

MAKES 4 SERVINGS

Preheat the oven to 400°F [200°C]. Line a rimmed baking sheet with foil.

In a large bowl, toss together **1 large sweet potato, peeled and diced into ¼ in [6 mm] pieces**; **1 Tbsp extra-virgin olive oil**; and **fine sea salt** to coat well. Cover with a sheet of foil to form a tight lid and roast for 20 minutes. Remove the top layer of foil and roast until the sweet potatoes turn golden brown and lightly caramelized, 15 to 20 minutes. Remove from the oven and let cool for 5 minutes.

In a large mixing bowl, massage with your hands **1 bunch dinosaur kale (about 10 oz [285 g]), midribs removed and discarded, leaves finely chopped fine**; **1 Tbsp extra-virgin olive oil**; and **¼ tsp fine sea salt** until the leaves start to soften and look wilted. Add the roasted sweet potatoes, **2 Tbsp roasted salted pumpkin seeds**, **2 Tbsp shredded Parmesan**, and **1 cup [120 g] crispy chickpeas, homemade (page 225) or store-bought**.

To make the Caesar dressing, in a small mixing bowl, smash **2 anchovy fillets packed in olive oil, drained, or 1 tsp anchovy paste**; **1 garlic clove, grated**; and **¼ tsp ground black pepper** until it forms a smooth paste. Fold in **½ cup [120 g] mayonnaise**, **1 tsp Dijon mustard**, **1 tsp Worcestershire sauce**, **½ tsp fish sauce**, **2 Tbsp grated Parmesan**, and **zest of 1 lemon**. Taste and season with **fine sea salt**. The dressing should have an intense umami flavor without tasting fishy; the cheese will predominate. Let it sit for at least 10 minutes and up to 30 minutes before using.

Toss the salad with 2 to 3 Tbsp of the dressing and serve the rest on the side. Serve immediately. Leftovers can be stored in an airtight container in the refrigerator for up to 3 days, but keep in mind the chickpeas will lose their crispiness as they're exposed to moisture.

After reading this book you might conclude that I really like Caesar salad dressing any which way. It's true. This version is the classic upon which Avocado Caesar Dressing (page 120) is based. Roasted sweet potatoes, tender massaged kale, and crispy chickpeas with delicious umami and salty notes are mingled with luscious creaminess.

- Roasted slices of delicata squash and pumpkin also work great as a substitute for sweet potatoes in this salad.

- Since this recipe uses hearty dinosaur kale (aka Tuscan kale or cavolo nero), the leaves need to be massaged to break down their cells. This helps make the leaves softer and easier to eat. You do not need to do this with baby kale because the leaves are already very tender.

- The anchovies can be replaced with 1 tsp white or yellow miso paste and 1 tsp liquid aminos.

Roasted Sweet Potatoes with Guajillo Chilli Salsa

MAKES 4 SERVINGS

Preheat the oven to 400°F [200°C]. Line a baking sheet with foil.

In a large bowl, toss together **2 large sweet potatoes, cut into 1 in [2.5 cm] chunks**; **2 Tbsp extra-virgin olive oil**; and **fine sea salt** to coat well. Spread in a single layer on the prepared baking sheet and roast for 25 to 30 minutes, until the sweet potatoes turn golden brown and are tender on the inside, rotating the baking sheet halfway through roasting. Remove and transfer to a serving plate.

Guajillo Chilli Salsa

While the sweet potatoes roast, prepare the salsa. In a small saucepan, bring to a rolling boil **1 cup [240 ml] water**; **2 dried guajillo chillies, stalks and seeds removed and discarded**; **1 Tbsp fresh oregano**; **1 Tbsp fresh thyme**; **1 tsp cumin seeds**; and **1 tsp whole black peppercorns**. Turn down the heat to low and simmer until the liquid is reduced to ½ cup [120 ml]. Transfer to a blender and rinse the pan.

Roast **1 large red bell pepper** and **1 peeled shallot** directly over the burner flame of a gas stove until the outsides begin to char. If you don't own a gas stove, sear the vegetables in a dry skillet or roast them in an oven by broiling on high, rotating often, until they start to blister all over, 7 to 12 minutes. Core the bell pepper and discard the stalk and seeds. Remove some but not all of the burnt skin. Add the bell pepper and shallot to the blender. Pulse on high speed until the mixture turns smooth and creamy. Transfer to the rinsed saucepan and cook over low heat until the mixture reduces in volume to about ¾ cup [180 g]. Remove from the heat, taste, and season with **1 tsp lime juice** and **fine sea salt**.

Top the roasted sweet potatoes with ⅓ cup [80 ml] of the salsa, then top with **2 Tbsp crème fraîche or sour cream**. It should melt and turn buttery. Garnish with **2 Tbsp chopped cilantro**.

Serve immediately with the extra salsa on the side. Leftovers can be stored in an airtight container in the refrigerator for up to 3 days.

Sweet potatoes are a vegetable that are best cooked by methods that bring out the sweetness but also play with the sugars' ability to caramelize. Enter roasting, a simple yet efficient technique that takes full advantage of the innate goodness hidden inside sweet potatoes and reveals their glorious flavor. This guajillo chilli salsa is served with a heap of crème fraîche that melts away to give a tangy, buttery taste. This salsa is also fantastic on its own when served with nachos, tortilla chips, tacos, and so on—you get the picture. You need to make it often.

THE COOK'S NOTES

- Guajillo chillies are considered a mild chilli. Larger varieties are milder in heat but richer in flavor than the smaller ones, which are slightly hotter. Dried chipotle chillies can also be used here, but they're hotter, so watch out!

- Leave the sweet potatoes unpeeled or peel them; it's up to you. I like the skin on.

- Boiling the dried guajillo chilli in water along with the spices helps hydrate and soften the fruit.

Kung Pao Sweet Potatoes

MAKES 4 SERVINGS

In a wok or large skillet over high heat, warm **2 Tbsp neutral oil with a high smoke point such as grapeseed**. When the oil begins to shimmer, add **1½ lb [680 g] orange-fleshed sweet potatoes, peeled and diced into ¼ in [6 mm] cubes,** and stir-fry until golden brown and tender, 7 to 9 minutes. Lower the heat if they start to turn too dark. Season with **fine sea salt** and **¼ tsp ground black pepper**. Transfer the sweet potatoes with a slotted spoon to a large plate or bowl.

In a small bowl, whisk together **¼ cup [60 ml] Shaoxing wine or dry sherry, 1 Tbsp low-sodium soy sauce, 2 tsp toasted sesame oil, 1 tsp Chinese black vinegar, 1 tsp cornstarch, 1 tsp ground Sichuan peppercorns,** and **½ tsp sugar** to form a smooth sauce.

Wipe out the wok or skillet and heat **2 Tbsp neutral oil with a high smoke point such as grapeseed** over high heat. When the oil begins to shimmer, add **10 to 12 whole dried red chillies such as chilli de árbol, stems and seeds removed and discarded, cut into ½ in [13 mm] pieces,** and stir-fry for 15 to 30 seconds, until fragrant and the chillies turn bright red and start to expand. Add **2 garlic cloves, thinly sliced,** and **2 in [5 cm] piece fresh ginger, peeled and cut lengthwise into thin slices,** and stir-fry for 30 seconds, until fragrant.

Return the sweet potatoes to the wok. Drizzle the sauce over the sweet potatoes and stir-fry until it thickens and the sweet potatoes are completely coated, 30 seconds. Add **½ cup [70 g] roasted unsalted whole peanuts** and **4 scallions, both white and green parts, cut into ½ in [13 mm] long pieces,** and stir for 1 minute. Remove from the heat, taste, and season with **fine sea salt**. Transfer to a serving bowl and serve immediately. Leftovers can be stored in an airtight container in the refrigerator for up to 3 days.

There are two foods that my husband, Michael, is extremely fond of: kung pao chicken and sweet potatoes. I've combined his love for both in this classic Chinese dish from the Sichuan province. Unlike the chicken, which is usually battered and fried, the sweet potatoes are stir-fried to bring out their sweet caramel flavor and then tossed with the sauce. Serve this warm with plain rice.

THE COOK'S NOTES

- The orange-fleshed sweet potatoes look much nicer in this dish, but the white ones also work.

- Sweet potatoes are rich in sugar, so you need to keep a watchful eye on them, as they can quickly burn and turn black and bitter.

- Shaoxing wine can be found online and in Asian grocery stores. A dry sherry is a good alternative.

- Chinese black vinegar, made with glutinous rice and malt, can be found online and in Asian markets and it's worth seeking out. Believe me, once you try it, you'll always want to keep it stocked in your pantry.

- Be vigilant and don't overcook the sweet potatoes and chillies; they both burn fast and will turn bitter.

Sesame Sweet Potatoes + Gochujang Chicken

Use a knife to make two or three gashes across the skin side and through the flesh of **4 to 6 chicken thighs, bone-in and skin-on (total weight about 2 lb [910 g])**. The cuts do not need to pass all the way through the chicken.

In a large bowl, combine ¼ **cup [60 ml] hot water**, **2 Tbsp gochujang paste**, **2 Tbsp grated fresh ginger**, **2 Tbsp apple cider vinegar**, **1 Tbsp honey**, **1 Tbsp neutral oil with a high smoke point such as grapeseed**, and ½ **tsp fine sea salt**. Mix until smooth. Add the chicken thighs and toss to coat. Refrigerate for at least 30 minutes and up to 1 hour to allow the flavor to infuse the chicken.

While the chicken marinates, preheat the oven to 400°F [200°C]. Line two rimmed baking sheets with foil, one for the chicken and the other for the sweet potatoes.

Place **2 medium sweet potatoes, scrubbed and cut in half lengthwise**, cut side up on one of the prepared baking sheets. Drizzle with **2 Tbsp sesame oil**. Sprinkle with **1 Tbsp sesame seeds**, ½ **tsp ground black pepper**, and **fine sea salt**. Cover with a second sheet of foil and fold the edges to form a tight seal. Roast for 30 minutes, remove the foil cover, rotate the pan halfway, and continue to roast, uncovered.

At this point, remove the chicken thighs from the marinade, letting the extra marinade drip off, then place them on the second baking sheet and roast on another rack.

Continue roasting the sweet potatoes until they are cooked thoroughly and tender (a knife should slide easily through the center), an additional 10 to 20 minutes. Remove the sweet potatoes from the oven and set them aside.

Roast the chicken until the skin is crisp and the internal temperature reaches 165°F [74°C] on an instant-read thermometer, for a total of 30 to 45 minutes. Let the chicken rest for 5 minutes. Before serving, garnish with **2 Tbsp chopped chives or scallion**. Serve immediately or while still warm.

This is a sheet pan meal, one that I turn to often not only during the week but also when we entertain. It's sweet and spicy, with gochujang and sesame doing all the work for you. This is one of those recipes that requires minimal effort and planning, and I file it under the category of "Minimal Work and the Cook Still Looks Ravishing" (which I wish also applied to my morning skin care routine).

THE COOK'S NOTES

• Gochujang, a staple in Korean cooking, is a gloriously red chilli paste made from fermented soybeans, glutinous rice, sweeteners, and salt. The degree of heat and sweetness vary by brand.

• This recipe works great on an outdoor grill, which gives a nice smoky, charred flavor.

• Because gochujang contains sugar, watch the chicken carefully as it cooks, or it could burn. If you're starting to see too much color, lower the temperature to 375°F [190°C] and adjust your cooking time as needed.

143

Cabbage
Bok Choy
Broccoli
Brussels Sprouts
Collards
Cauliflower
Romanesco
Radishes
Arugula
Kale
Mustard Greens
+ Watercress

Origins

CABBAGE HAILS FROM SOUTHERN AND WESTERN EUROPE. BOK CHOY AND RADISH ARE ORIGINALLY FROM CHINA. BROCCOLI AND CAULIFLOWER COME FROM THE EASTERN MEDITERRANEAN AND ASIA MINOR. BRUSSELS SPROUTS ARE FROM THE MEDITERRANEAN AND NORTHERN EUROPE, AND COLLARDS COME FROM GREECE. WHITE AND YELLOW MUSTARDS ARE FROM THE MEDITERRANEAN, AND BROWN MUSTARD IS FROM THE HIMALAYAS. KALE HAILS FROM EUROPE AND ASIA MINOR, WATERCRESS FROM EUROPE AND ASIA, AND ARUGULA FROM THE MEDITERRANEAN AND SOUTHERN EUROPE.

Cabbage

One of the few vegetables that I really looked forward to eating all the time as a kid, cabbages are orbs of love. Cabbages come in a variety of sizes, shapes, and textures, and in green and purple-red colors. Whatever way it's prepared, cabbage always tastes good. Fermented cabbage in pickles and kimchi (napa cabbage is used for kimchi) are delicious and a wonderful shortcut to build flavor in a dish (see Kimchi Creamed Corn, page 73). Use the outer damaged cabbage leaves to line steamers when making dumplings or steaming vegetables (see Collards Patra, page 148, or Bok Choy and Crispy Tofu, page 166).

Bok Choy

Bok choy, or pak choy, is a type of Chinese cabbage and is sold in both large and baby sizes. The leaves and stalks are both eaten. Stir-frying is my favorite way to eat them. If the stalks are tough, blanch them first before stir-frying (see Bok Choy and Crispy Tofu, page 166).

Broccoli and Broccolini

The edible part of broccoli includes the flower head, the stalk, and the small, attached leaves. Occasionally, the flower head blooms into tiny yellow flowers after harvesting, especially when stored near fruits like apples that produce ethylene, a plant-ripening hormone. It's still edible.

Broccolini, or baby broccoli, is a hybrid of broccoli and gai lan (Chinese broccoli). It was developed by the Sakata Seed Corporation in Japan. Both broccoli and broccolini can be cooked by the same methods of stir-frying, steaming, boiling, roasting, and grilling. Rapini, or broccoli rabe, is not a type of broccoli but a variety of turnip, another member of the brassica family (in case you're wondering about the absence of turnips in this section, they're one of the few vegetables I avoid, as I dislike their smell).

Brussels Sprouts

Brussels sprouts—or "little cabbages" as I playfully refer to them—are best either roasted or thinly shaved for raw salads. Boiled Brussels sprouts are a terrible idea and do a disservice to this vegetable. When purchasing, opt for smaller Brussels sprouts over larger ones as they taste sweet rather than sulfuric.

Collards or Collard Greens

Unlike some of the other brassica family members like cabbages and bok choy, collards are a loose-leafed variety of plants. In Kashmir, collards are cooked with mustard oil, dried red chillies, and asafetida to make a fragrant dish called haak.

Cauliflower

Cauliflowers are flower heads with tightly packed tiny primordial buds that never flower due to a genetic mutation. This head is sometimes called a *curd* and, when cut in half, it always reminds me of the human brain. Cauliflower comes in white, purple, yellow, orange, and green and the outer leaves resemble collard greens. Cauliflower can be eaten whole, chopped, or grated, and the latter is great in stuffed parathas and is used as a grain-free "rice" option.

Romanesco

This geometric marvel of the brassica world is often erroneously described as a hybrid between cauliflower and broccoli, but it is actually a variety of cauliflower that developed a genetic mutation in two genes involved in flower production that creates the unique fractal pattern of repeating spiraling cones. A fun nerdy fact: The number of spirals correlates with the numbers in the Fibonacci sequence (a series of numbers in which each number is the sum of the two that precede it, for example 0, 1, 1, 2, 3, 5, and so on). Romanesco comes in different shades of white, yellow, orange, green, and purple. This flower head goes by various names: Romanesco broccoli, Romanesco cauliflower, Roman cauliflower, and fractal broccoli. When cooking, I treat it like cauliflower, but with one exception:

I rarely grate it, because that would do a disservice to this fractal beauty. Use it in place of the cauliflower head in the Royal Cauliflower Roast with Almond Cream (page 169).

Radish

Radishes are the perfect source of crunchiness in salads and sandwiches. They come in a gorgeous assortment of colors, and they make my heart jump with joy when I see them in a salad. Radishes are sorted into summer and winter varieties depending on their life cycle: The smaller cherry, oblong, and elongated varieties like Easter Egg radishes are summer radishes, while the larger round and oblong varieties like daikon belong to the winter group. The black radish is one of the spiciest varieties and can be eaten raw or roasted at 425°F [220°C] for a milder flavor. The mildest tasting radishes are the smaller cherry-shaped varieties such as French Breakfast. These are wonderful when dipped in butter and salt but also fantastic paired with dips like Peanut Muhammara (page 275) or Pumpkin Seed Chutney (page 190).

Arugula, Kale, Mustard Greens, and Watercress

I've grouped this crew of greens because of the way we use them. They are wonderful in salads and sandwiches or as a garnish or topping. Arugula, kale, and mustard greens can be eaten both raw and cooked (see Beets, Toasted Barley + Burrata Salad, page 103), but watercress shines brilliantly when eaten raw.

Storage

Cabbage is one of the hardiest vegetables to store; it seems to last forever. Broccoli and broccolini benefit from being stored in open plastic bags in the refrigerator;

otherwise, they overripen and yellow quickly. Wrap unwashed arugula, kale, mustard greens, and watercress in paper towels before storing them in the refrigerator.

Cooking Tips

- When purchasing most green vegetables, generally avoid any that are yellowing or spotted with black. Yellowing is an indicator of overripening, a sign that the bright green chlorophyll pigment is waning and that fresh green is on its way out.

- When brassicas are cut or chopped, their broken cells release an enzyme called *myrosinase* that produces a sulfurous smell and bitter-tasting substances. This ability developed as a defense mechanism to protect the plant from animals and insects. We humans, on the other hand, either enjoy the flavor—or don't.

- The sulfurous flavor of freshly cut brassicas such as cabbage, Brussels sprouts, and radishes in raw dishes like salads can be mellowed by submerging them in a bowl of ice-cold water for 15 to 30 minutes after chopping. Drain the water and pat the vegetables dry before using. The cold water prevents the enzyme responsible for creating that sulfurous smell from doing its job and washes away any of these flavors created during chopping. For preparations that involve heat, high temperatures destroy the enzyme, so there is no need to soak the vegetables before cooking.

- Contrary to popular belief, lemon juice does not prevent myrosinase from producing sulfurous smells. The vitamin C (ascorbic acid) present in lemon and other citrus juices actually acts as a cofactor and helps accelerate the biochemical reaction. I've noticed that vinegar doesn't help much either—pickled brassicas can smell quite potent.

- Pairing wines with brassicas is a little tricky, as the sulfurous flavors of brassicas interfere with the appreciation of wine's subtle complexity. In general, white wines tend to fare a little better; red wine and champagne aren't the best options.

- Buy broccoli and broccolini that's firm and not dry. Trim the tough bottom ends before cooking. If the stalks turn limp, trim the base and submerge it in a cup of water, cut side down, and refrigerate for a few hours to help revive them.

- Strip the midribs and the tough end of the stalks off large-leafed varieties like dinosaur kale. This is unnecessary with tender baby kale.

- Cabbage leaves are sturdy, which makes them great for stuffing and forming rolls. However, you must soften them slightly to make them pliable enough to fold without ripping. Dip them in boiling water, as is done in the recipe for Stuffed Cabbage Rolls in Tomato Sauce (page 174).

- If you can't find watercress, baby arugula is a good substitute. Massaging chopped or whole kale leaves with olive oil and salt uses mechanical action to break down the tough cellular fibers and improve the mouthfeel (see Sweet Potato Kale Caesar Salad, page 136). Massaging benefits the tougher varieties of kale, such as dinosaur (which is also called lacinato, Tuscan, or cavolo nero). There's no need to massage baby kale; it is already tender.

147

Collards Patra

Wash and pat dry **1 bunch large collard green leaves (about 5 to 8 large leaves, total weight about 9 oz [260 g])**. Cut out and discard the thickest part of the midrib and the stalks.

In a medium bowl, combine **1½ cups [180 g] sifted chickpea flour, 1½ tsp fine sea salt, 1 tsp ground coriander, ½ tsp smoked sweet paprika, ½ tsp ground turmeric,** and **¼ tsp ground cayenne**.

Prepare the yogurt coating. In a separate medium bowl, combine **½ cup [120 g] plain unsweetened yogurt, ½ cup [120 ml] water, 2 Tbsp extra-virgin olive oil, 1 Tbsp tamarind paste,** and **1 Tbsp peeled and grated fresh ginger**. Whisk until smooth. Fold the yogurt mixture into the chickpea flour mixture to form a smooth, thick paste. There should be no clumps.

Lay one leaf down flat with its glossy side up and the pointy top of the leaf pointing away from you. Using a pastry brush or an offset spatula, spread a thin layer of the chickpea-yogurt mixture—about 1½ Tbsp or just enough to coat the entire surface of the leaf. Place the second leaf over the first leaf, this time with the pointy tip of the leaf pointing toward you, and coat it with the chickpea-yogurt mixture. Repeat with the remaining leaves, alternating in the same way.

Now get ready to fold the stack of leaves like a burrito. Start by gently pressing down about 1 in [2.5 cm] from the longer left and right sides inward while you roll the stack of leaves from the top as tightly as you can. Coat the surface of the folded edges as you tightly roll it inward to form a log. Cut the log into 2 equal parts to form rolls, using a sharp serrated knife and a sawing motion. The logs won't unfold, but if you're worried, you can stick a wooden toothpick through the roll to hold it together.

Set up a steamer and fill the base with about 1 in [2.5 cm] of water. At no point should the base of the steamer touch the water. Line the base of the bamboo steamer with parchment paper or lettuce leaves. If using a metal steamer, grease the insides with a little vegetable oil. Bring the water to a boil over medium-high heat, then lower the heat to a simmer. Place the rolls in the steamer basket. Cover and steam until the leaves turn a dark shade of green and the chickpea-yogurt mixture firms up, 9 to 10 minutes.

Remove from the basket and transfer to a plate. Cut each log into 4 equal, spiral pieces (you will have a total of 8).

In a 12 in [30.5 cm] cast-iron or stainless-steel skillet heat **2 Tbsp neutral oil with a high smoke point such as grapeseed** over medium heat. Add the steamed rolls and fry on each side until they turn golden brown and crisp, 2 to 3 minutes per side. Remove from the heat and transfer to a plate.

Wipe the skillet clean and heat **2 Tbsp neutral oil with a high smoke point such as grapeseed** over medium-high heat. When the oil is hot, add **1 Tbsp whole black or brown mustard seeds** and **12 to 15 fresh curry leaves**. Set the crisped rolls on the spices in the hot oil and press down gently for 30 seconds. Then flip the rolls and fry on the opposite side for another 30 seconds, until the spices are fragrant and the curry leaves turn crisp and translucent.

Remove and transfer the rolls and curry leaves to a serving plate. Garnish with **2 Tbsp chopped cilantro**; **1 Tbsp fresh grated coconut (optional)**; and **1 fresh green or red chilli such as serrano, jalapeño, or bird's eye, minced**.

Serve warm or at room temperature. Patra is best eaten the day it's made, but you can reheat it in a 300°F [150°C] oven until warm. Leftovers can be stored in an airtight container in the refrigerator for up to 2 days.

Patra or alu vadi comes to us by way of the western state of Gujarat in India. The dish is typically made with the large, elephant-earlike leaves of the taro plant, but since those are hard to come by, I started to make them with collard leaves. While these are smaller than taro, they are similar in texture and amenable to folding. The process to prepare these swirled rolls is simple. Lay a leaf down, coat it with the chickpea-yogurt mixture, lay another leaf down, and repeat. The entire stack of leaves is then rolled to form a burrito-like log that's sliced and pan-fried until crisp. Serve this warm as a snack or as a side to a meal.

THE COOK'S NOTES

- You will need a steamer. Either a bamboo or a metal steamer that fits into a wok or pot will work.

- Select large leaves with minimal holes and tears. Some slight tears will be covered during rolling, but it's best to minimize leakage of the yogurt mixture.

- Do not use Greek yogurt here; it is too thick. In fact, if necessary you can thin the yogurt with a little cool water.

- The assembly process will remind you of rolling up a burrito or a jelly-roll cake. If the roll won't hold its shape, tie it with two pieces of kitchen string. Remove the strings after steaming.

continued

THE COOK'S NOTES (CONTINUED)

- Tamarind paste (also called purée) is sold in grocery stores in the Asian and Indian aisles, or in international markets. Tamarind paste in jars should be a syrupy consistency.

- To be on the safe side, I've given you amounts to make a bit more of the yogurt mixture than you need because leaf sizes will vary. It's not likely you'll need all of it. Avoid thickly coating the leaves just to use up the mixture; keep the layers thin. Thicker layers of the yogurt produce a cakier texture and the patra won't be as crisp.

- Patra are cooked twice. The rolls are first steamed to help the starch in the chickpea flour gelatinize and form a seal between the leaves. The steamed rolls are then lightly pan-fried until crispy.

- The spices are usually added to the frying step, but I prefer to add them in stages. This prevents the spices from burning and turning bitter because the cooking times for the patra and the mustard seeds are significantly different.

8. Cabbage, Bok Choy, Broccoli, Brussels Sprouts, Collards, Cauliflower, Romanesco, Radishes, Arugula, Kale, Mustard Greens + Watercress

Brassica Fritters, Okonomiyaki Style

MAKES 4 PANCAKES

Preheat the oven to 300°F [150°C]. Line a plate with paper towels.

Prepare the dipping sauce by combining in a small mixing bowl **¼ cup [65 g] ketchup**, **¼ cup [60 ml] Worcestershire sauce**, and **1 tsp low-sodium soy sauce**. Whisk until smooth.

In a large mixing bowl, whisk together **1 cup [120 g] store-bought tempura batter mix**, **1 tsp dashi stock powder**, **¼ cup [35 g] rice flour**, **¼ cup [35 g] sesame seeds**, **1 tsp baking powder**, and **¼ to ½ tsp fine sea salt**. Stir in **1 cup [240 ml] ice-cold water** and whisk until smooth, with no lumps or flecks of dry flour. Fold in **1 cup packed finely shredded green cabbage (about 7 oz [200 g])**; **6 Brussels sprouts, thinly shaved (about 4¼ oz [120 g])**; **4 scallions, both white and green parts, thinly sliced**; and **1 sheet of nori, cut into thin 4 in [10 cm] long strips**.

In a small cast-iron skillet or frying pan over medium-high heat, warm **1 Tbsp neutral oil with a high smoke point such as grapeseed**. Drop ½ cup of the batter and spread with the bottom of the measuring cup or swirl the skillet to form a rough 6 in [15 cm] circle. Cook until golden brown on each side, about 3 minutes per side, then transfer to the prepared plate to drain. Cook the remaining pancakes, adding more oil as needed between batches. As you work, keep the prepared pancakes warm, covered with foil, in the oven. Serve warm with the sauce on the side. These are best eaten fresh and hot out of the skillet. Leftovers can be stored in an airtight container in the refrigerator for up to 2 days.

This is not the classic okonomiyaki, the beloved savory pancake from Japan. By my interpretation, it's more of a fritter than a pancake. This recipe came to be during those early days of the pandemic when essentials like eggs quickly went into short supply at grocery stores and we couldn't eat out at restaurants. This is now a breakfast and lunch staple at our home.

THE COOK'S NOTES

- I like the Kikkoman brand of tempura batter mix; look for the "extra crispy" variety.

- Use regular rice flour, not glutinous rice flour.

- The ice-cold water is essential to achieving a crispier texture. It prevents the starch from gelatinizing, which in turn allows the water to evaporate much faster from the batter during cooking and produces a crispy texture.

- Thinly shaved vegetables are the key to everything cooking uniformly. A mandoline (wear Kevlar gloves) or Y-shaped vegetable peeler is your friend.

8. Cabbage, Bok Choy, Broccoli, Brussels Sprouts, Collards, Cauliflower, Romanesco, Radishes, Arugula, Kale, Mustard Greens + Watercress

Cabbage with Date + Tamarind Chutney

MAKES 4 SERVINGS

In a wok, large saucepan, or Dutch oven, warm **2 Tbsp extra-virgin olive oil** over medium-high heat. When the oil is hot, add **1 tsp black or brown mustard seeds** and **12 to 15 fresh curry leaves** and cook until fragrant, the seeds start to pop, and the leaves start to turn translucent and crisp, 30 to 45 seconds.

Turn down the heat to low and fold in **1 large green cabbage, cored and shredded**; **¼ cup [50 g] red lentils, rinsed and drained**; **¼ cup [60 ml] water**; **1 tsp ground black pepper**; and **fine sea salt**. Cover with a lid and cook until the cabbage and lentils are completely tender, 25 to 30 minutes. Stir occasionally to prevent burning.

Remove from the heat, transfer to a serving bowl, and garnish with **2 Tbsp chopped cilantro**, and **1 fresh chilli such as jalapeño, serrano, or bird's eye, thinly sliced**.

While the cabbage cooks, prepare the tamarind chutney. In a small saucepan over medium heat, whisk together **2 Tbsp water**; **2 Tbsp date syrup**; **1 Tbsp tamarind paste**; **1 Tbsp peeled and grated fresh ginger**; **½ tsp red pepper flakes such as Aleppo, Maras, or Urfa**; and **½ tsp ground cumin**. Bring to a boil and remove from the heat. Taste and season with **fine sea salt**.

Serve the cabbage-lentil mixture immediately, drizzled with 2 Tbsp of the date and tamarind chutney. Serve the rest on the side. Store leftovers in an airtight container in the refrigerator for up to 3 days.

When I was growing up, my favorite way to eat cabbage was a dish that my maternal grandmother, Lucy, prepared with lentils. I looked forward to it every time we visited her home. This recipe is inspired by her recipe, and I've added the date and tamarind chutney for a pop of sweet and sour goodness.

THE COOK'S NOTES

- Because the cabbage is cooked here, I suggest cutting it into wider shreds with a chef's knife. If finely shredded, it will release too much liquid and the texture will suffer.

- Fresh curry leaves can be purchased at your local Indian and Asian grocery stores.

- Tamarind paste is sometimes called tamarind concentrate or pulp, depending on the brand. What you're looking for is a thick liquid made from water and tamarind fruit flesh.

Radish Salad with Black Vinegar

MAKES 4 SERVINGS

In a small bowl or jar, whisk or shake together **¼ cup [60 ml] sesame oil or extra-virgin olive oil**, **3 Tbsp Chinese black vinegar**, **2 Tbsp minced preserved lemon peel**, **½ tsp ground black pepper**, and **¼ tsp ground coriander**. Taste and season with **fine sea salt**. The vinaigrette can be made at least 2 days in advance.

Separate the leaves from the roots of **1 bunch (10¼ oz [290 g]) radishes**. Discard the stems. Slice the radishes into thin disks and place them in a medium bowl. Rinse the leaves well, then chop and add them to the bowl.

In a stainless-steel skillet over medium-high heat, toast **2 Tbsp raw, unsalted pumpkin seeds** until the seeds start to turn light golden brown, 1½ to 2 minutes.

Pour some of the vinaigrette over the radishes. Add the pumpkin seeds. Toss to coat well, add more vinaigrette, and season with **fine sea salt** to taste. Serve within 2 hours of mixing, or the vinegar will soften the crispness of the radishes.

This cool, refreshing salad uses both the radish leaves and roots. Crisp, thin radish slices are dressed in a vinaigrette made with black vinegar, a common staple in Chinese and Taiwanese cuisines, and the preserved lemons that are so important in Middle Eastern pantries.

THE COOK'S NOTES

- Use only the peel of the preserved lemons in this recipe. (See how to use preserved lemons on page 94.)

- For a stronger sesame flavor, use toasted sesame oil in the vinaigrette.

- Black vinegar is extremely fragrant. It's made by fermenting a steamed grain like rice, sorghum, or wheat. The brand I use at home is Kong Yen, which is available online as well as in Asian grocery stores. Malt vinegar is a close alternative.

- You can toast your coriander seeds before incorporating them into the vinaigrette. To make a small batch, toast 2 Tbsp whole coriander seeds over medium heat in a dry, stainless-steel skillet until fragrant and lightly browned. Transfer the seeds to a cool plate. Once the seeds are cooled, grind them into a fine powder, store in an airtight container at room temperature, and use as needed for up to 1 month.

159

Broccoli Za'atar Salad

MAKES 4 TO 6 SERVINGS

Prepare the dressing by combining in a small bowl **¼ cup [60 ml] extra-virgin olive oil**; **¼ cup [60 ml] date syrup**; **3 Tbsp pomegranate molasses**; **1½ Tbsp za'atar, homemade (page 341) or store-bought**; **1 tsp red pepper flakes such as Aleppo, Maras, or Urfa**; and **1 tsp fine sea salt**.

In a large mixing bowl, combine **2 lb [910 g] broccoli florets, separated into small bite-size pieces**; **1 small red onion, minced**; **¼ cup [35 g] golden raisins**; **¼ cup [35 g] dried sweetened cranberries or dried sweet and tart cherries**; **¼ cup [30 g] toasted sliced almonds**; **2 Tbsp roasted unsalted pumpkin or sunflower seeds**; and **2 Tbsp chopped cilantro or flat-leaf parsley**.

Drizzle the dressing over the broccoli in the bowl. Toss to coat well. Taste and season with **fine sea salt**. Serve immediately. Leftovers can be stored in an airtight container in the refrigerator for up to 3 days.

This is a fruitier and spicier rendition of the broccoli salad seen in the grocery store deli, also sans the mayo. Aleppo pepper, za'atar, date syrup, and pomegranate molasses bring their game to this salad party.

THE COOK'S NOTES

- If the broccoli stalks are too tough, don't add them to the salad, but save them to add elsewhere or use in stock. Tender stalks can be thinly sliced and added to the salad.

- Roasted broccoli and broccolini also work great here.

Roasted Fruit + Arugula Salad

MAKES 4 SERVINGS

Preheat the broiler.

In a large bowl, toss to combine **1 lb [455 g] destemmed grapes** (any kind—a colorful mix is always wonderful to look at on a plate); **8 large ripe figs (about 7 oz [200 g]), cut in half lengthwise**; **2 Tbsp extra-virgin olive oil**; **1 Tbsp mix of black and white sesame seeds**; **1 Tbsp poppy seeds**; **½ tsp ground black pepper**; and **½ tsp fine sea salt**.

Spread the mixture out in a single layer in a heatproof 9 by 13 in [23 by 33 cm] rectangular roasting or baking dish. Make sure the cut side of the figs face upward. Transfer to the oven and broil on high, about 6 in [15 cm] from the heat source, until the fruit starts to turn golden brown and the grape skins split, 5 to 6 minutes. Remove the dish from the oven and let sit for 5 minutes.

Fold in **2 cups [40 g] tightly packed baby arugula or watercress**; **1 tsp dried ground oregano, preferably Mexican**; and **½ tsp dried ground thyme**. Tuck an **8 oz [230 g] block feta, sliced into 8 pieces each about ¼ in [6 mm] thick**, underneath the fruit and arugula. Drizzle with **2 Tbsp balsamic vinegar** and top with the **zest of 1 orange**.

Serve warm. Leftovers do not store well, so eat this the day it's made, preferably within an hour of preparing it.

The fresh fruit lover in me adores this salad. Gooseberries and stone fruit like peaches, apricots, and nectarines are also good choices, but to be honest, you can use any kind of fruit that can withstand roasting, so use what's in season and easy to get your hands on. The poppy and sesame seed combination adds delightful texture, and the hint of orange zest at the end is a sweet kiss.

THE COOK'S NOTES

- Choose a nice-looking vessel to roast the fruit in, so you can take it directly from the oven to the table.

- When broiling the fruit, pay careful attention to avoid burning. Fruits are rich in sugar and can char easily.

- If you have Mexican oregano, use it here; it's a little spicier than some of the other varieties and makes the flavor pop.

- Grilled slices of kefalotyri also works great in place of feta.

8. Cabbage, Bok Choy, Broccoli, Brussels Sprouts, Collards, Cauliflower, Romanesco, Radishes, Arugula, Kale, Mustard Greens + Watercress

Sweet + Sticky Brussels Sprouts

MAKES 4 SERVINGS

Preheat the oven to 400°F [200°C].

In a medium saucepan, combine **2½ cups [600 ml] water**; **1 cup [200 g] black or forbidden rice, rinsed and drained**; and **fine sea salt**. Bring to a boil over medium-high heat, lower the heat to a simmer, cover, and cook until the water is completely absorbed by the rice and the grains turn tender, 40 to 50 minutes. Remove from the heat and let rest, covered, for 15 minutes.

After the rice has been cooking for about 30 minutes, toss on a sheet pan or baking dish **1½ lb [680 g] Brussels sprouts, cut in half lengthwise**; **2 Tbsp sesame oil**; and **fine sea salt**. Spread the sprouts out in a single layer. Roast until golden brown and slightly charred, 22 to 30 minutes, stirring halfway through. Transfer to a large mixing bowl.

While the sprouts roast, prepare the sauce. In a small saucepan, whisk together **¼ cup [60 ml] mirin**, **3 Tbsp honey or maple syrup**, **2 Tbsp white or yellow miso paste**, **1 Tbsp rice wine vinegar**, **1 Tbsp sesame oil**, **2 tsp low-sodium soy sauce**, and **½ tsp ground black pepper**. Bring to a boil over medium-high heat, lower the heat to a simmer, and cook until the sauce starts to thicken, 30 to 60 seconds. Remove from the heat, taste, and season with **fine sea salt**.

Pour the sauce over the hot Brussels sprouts in the bowl and fold to coat well. Serve the Brussels sprouts immediately with the warm rice, garnished with **2 scallions, both white and green parts, thinly sliced**; **2 Tbsp whole cilantro leaves**; and **1 Tbsp toasted sesame seeds**.

There is a general and universally accepted rule that the best way to prepare Brussels sprouts is roasting, not boiling. I stick to that edict. Roasting caramelizes Brussels sprouts and makes their leaves delightfully crispy. The finishing touch is the sweet and sticky miso-based sauce. It's important to add the sauce to *hot* Brussels sprouts— the sauce sticks much better, and you'll hear the vegetables sizzle. Pair this dish with the Radish Salad with Black Vinegar (page 159).

THE COOK'S NOTES

- Toasted sesame seeds are available at most grocery stores, but you can make your own. Heat white sesame seeds in a small dry skillet over medium-low heat, swirling the pan occasionally, until they turn fragrant and light golden brown.

- I don't use toasted sesame oil to cook the Brussels sprouts; its smoke point is much lower than regular sesame oil. If you want a stronger sesame flavor, drizzle 1 to 2 Tbsp of toasted sesame oil over the dish just before serving.

Bok Choy with Crispy Tofu

MAKES 4 SERVINGS

In a medium saucepan over medium-high heat, bring to a boil **2½ cups [600 ml] water**; **1 cup [200 g] black or forbidden rice, rinsed and drained**; and **fine sea salt**. Turn down the heat to low, cover, and let simmer until all the water is absorbed by the grains and they turn tender, 40 to 50 minutes. Remove from the heat and let sit, covered, for 15 minutes.

In a large mixing bowl, toss together **½ cup [30 g] panko**, **½ cup [70 g] sesame seeds (use a mix of black and white)**, **½ tsp ground black pepper**, and **fine sea salt**.

In a medium bowl, whisk **2 large eggs**.

Using clean paper towels or a lint-free kitchen towel, pat dry one **12 oz [340 g] package extra-firm tofu, cut into eight rectangles**. Using a pair of kitchen tongs or a fork, dip the tofu slices one at a time in the beaten eggs. Tap gently on the side to drain off any excess liquid and then transfer to the bowl with the panko mixture. Toss gently to coat well and place on a plate. Repeat with the remaining tofu slices.

Line a plate with paper towels. In a cast-iron or stainless-steel skillet over medium heat, warm **¼ cup [60 ml] neutral oil with a high smoke point such as grapeseed**. Working in batches to avoid crowding, panfry the breaded tofu slices in the hot oil on each side until crisp and golden brown, 1½ to 2 minutes per side. Repeat with the remaining tofu, adding more oil and wiping out the pan as necessary. Place the tofu slices on the paper towels to soak up any excess oil.

Fill a steamer with 1 in [2.5 cm] of water and bring to a boil over high heat. Place in the steamer basket **1 lb [455 g] baby bok choy (4 or 5 heads), stalks trimmed and leaves separated**. Steam for 1 minute. Remove from the steamer and rinse under cool running water to stop the cooking as well as flush away any dirt from between the leaves. Drain and pat dry with clean paper towels or a lint-free kitchen towel.

Wipe the skillet clean. Warm over high heat **2 Tbsp neutral oil with a high smoke point such as grapeseed**. When the oil shimmers, add **1 large onion, halved and cut into thin slices**. Stir-fry until it just starts to turn translucent, 4 to 5 minutes. Add **2 garlic cloves, thinly sliced**, and stir-fry until fragrant, 30 to 45 seconds.

continued

Add the steamed bok choy and stir-fry until the stalks just start to brown, 1½ to 2 minutes. Remove from the heat and let cool for 1 to 2 minutes (this ensures that the vinegar added next won't be boiled off by the hot skillet).

Drizzle with **2 Tbsp rice vinegar or Chinese black vinegar** and **1 Tbsp low-sodium soy sauce**. Taste and season with fine sea salt.

To serve, transfer the rice to a serving bowl and top with the cooked bok choy and fried tofu.

Garnish with **2 scallions, both white and green parts, thinly sliced**, and top generously with **2 Tbsp chili crisp** (and use as much as your heart desires). Serve immediately. Leftovers can be stored in an airtight container in the refrigerator for up to 3 days.

Based on one of my most popular recipes for the *New York Times Cooking Section*, this is one of my favorite ways to cook tofu; crisp outside and tender inside. And based on the comments I've received from people, it has converted many tofu-averse people into tofu lovers. Tender leaves of baby bok choy and a heap of chili crisp oil make the crispy tofu's texture shine with bold flavors.

THE COOK'S NOTES

- Watch the tofu carefully during cooking; if the sesame seeds burn, they will taste bitter.

- Rinse the bok choy leaves well to remove any sand or dirt that might be trapped in between them.

- It might seem like a tiresome extra step, but steaming the bok choy before it goes into the wok ensures a bright-green color, the thicker part of the stalks cooks evenly, and the vegetable doesn't turn mushy.

Royal Cauliflower Roast with Almond Cream

MAKES 4 SERVINGS

In a medium heatproof bowl, cover **1 cup [140 g] raw whole almonds** with enough boiling water to completely submerge them. Cover and let sit for 30 minutes. Drain and discard the water. Rub the skins off the almonds and discard. Transfer the almonds to a blender.

Add **1 cup [240 ml] low-sodium vegetable stock or Master Mushroom Vegetable Stock (page 337)**; **2 Tbsp lemon juice**; **1 Tbsp maple syrup**; and **1 tsp poppy seeds**. Pulse over high speed for 30 to 60 seconds, until smooth and creamy. Taste and season with **fine sea salt**. Transfer to a small serving bowl. The sauce can be made a day in advance and stored in an airtight container in the refrigerator.

Preheat the oven to 425°F [220°C]. Line a large roasting pan or baking sheet with foil or a wire rack.

Fill a large stockpot with enough water to eventually cover **1 large cauliflower**. Stir in **fine sea salt** (see the Cook's Notes), cover the pot with a lid, and bring to a rolling boil over high heat.

Trim and discard most of the outer leaves from the cauliflower and trim the basal stalk to make sure the cauliflower can sit comfortably upright. Invert and submerge the cauliflower, flower side down, in the boiling water. Blanch for 5 to 6 minutes, until it starts to just turn tender and a little translucent. A knife or skewer should pass through the cauliflower without much resistance. Carefully remove the cauliflower from the stockpot using a pair of kitchen tongs and a perforated ladle. Let the cauliflower sit, upright and flower side up, on a cutting board to drain off any excess water, and then place it on the roasting pan.

While the oven and water heat up, prepare the seasoning mixture. In a small bowl, combine **3 Tbsp extra-virgin olive oil, unsalted melted butter, or ghee**; **1 tsp cumin seeds**; **1 tsp fennel seeds**; **1 tsp poppy seeds**; **1 tsp chia seeds**; **1 tsp ground black pepper**; **1 tsp ground turmeric**; **1 tsp Kashmiri chilli powder (or ¾ tsp smoked sweet paprika powder + ¼ tsp ground cayenne**. Brush the mixture all over the cauliflower. Season with **fine sea salt**. Roast the cauliflower until the surface turns golden brown on the outside, 15 to 20 minutes, rotating halfway through during the roasting time.

continued

Remove from the oven, let rest uncovered for 5 minutes, and transfer to a serving platter. Garnish with **2 Tbsp chopped cilantro leaves** and **1 fresh green or red chilli such as bird's eye, jalapeño, or serrano, minced**. Serve immediately with the almond cream on the side and a carving knife. Leftovers can be stored in an airtight container for up to 3 days.

This is an elegant entrée based on the creamy shahi (which means "royal") dishes of the Moghul empire. It commands attention on a swanky dinner table. Bring it out after all the guests are seated, and use your best serving platter to present it as dramatically as possible.

THE COOK'S NOTES

- Blanch your own almonds or buy them blanched. Either way, the skins must be removed, or the sauce won't taste as nice.

- Be careful when maneuvering the cauliflower in the hot water; don't be too rough with it, as you risk damaging the florets.

- If you own a rotating cake stand, use it here. Place the cauliflower on it and rotate while you brush on the sauce.

- Everyone salts their water differently when boiling vegetables or pasta. I use 1 tsp fine sea salt for every 4½ cups [1 L] water. Adjust the amount of salt depending on the volume of water used. The salt helps with tenderizing the vegetables, but it also gets into those tiny nooks and crannies inside the cauliflower, helping with seasoning.

- Romanesco is a lovely and quite dramatic alternative to cauliflower.

170

Stuffed Cabbage Rolls in Tomato Sauce

MAKES 14 TO 16 ROLLS

Set up two saucepans, large and medium. To the large saucepan, add **1¼ lb [570 g] russet potatoes, cut into 2 in [5 cm] pieces**, and add enough water to completely cover the potatoes. Stir in **fine sea salt** (see the Cook's Notes). Cover with a lid and bring to a rolling boil over high heat, then lower the heat to a simmer and cook until the potatoes turn tender and a knife or skewer passes through the center with minimal resistance, 20 to 30 minutes. Remove the potatoes with a slotted spoon and transfer to a large mixing bowl to cool. Once cool enough to handle, peel and discard the skin. Mash the potatoes with a large fork or masher.

To the medium saucepan, add **½ cup [100 g] beluga lentils, picked over and rinsed**; **2 cups [480 ml] water**; and **1 tsp fine sea salt**. Bring to a boil over high heat, then lower the heat to a simmer and cook until the lentils are completely tender but aren't falling apart, 25 to 30 minutes. Drain the lentils through a fine mesh sieve set over the sink and add to the mashed potatoes.

Fold in **1 green chilli such as jalapeño or serrano, minced, deseeded for a milder heat**; **2 Tbsp chopped cilantro**; **2 tsp garam masala, homemade (page 341) or store-bought**; **1 tsp Kashmiri chilli powder (or ¼ tsp smoked sweet paprika + ¼ tsp ground cayenne)**; and **½ tsp ground black pepper**. Taste and season with **fine sea salt**.

Line a baking sheet with parchment paper or wax paper to prevent sticking. Divide the mixture into 14 to 16 equal parts, about 3 Tbsp each, and shape each into a 4 in [10 cm] cigar-shaped log. Arrange them on the prepared pan.

Separate **14 to 16 large leaves** from **1 cabbage head, preferably Savoy (you might need 2 Savoy cabbages)**. Work with one leaf at a time. Lay one leaf on a cutting board, with the curved side facing upward. Using a small paring knife, make a small, narrow V-shaped cut at the bottom of the leaf to remove the tough white stalk. This helps the leaves fold and hold their shape after rolling. Repeat with the rest of the leaves.

Set up a large saucepan filled with enough salted water to cover several leaves at a time and bring to a boil over high heat. Submerge four cabbage leaves at a time and blanch for 1 minute, until tender and translucent. Transfer the leaves with a slotted spoon to a cutting board. Repeat with the remaining leaves. Once the leaves are warm but cool enough to handle, lay a leaf flat on the cutting board, with the wide end of the notch toward you. Place one log of the filling near the tip of the notch. Fold the sides of the leaf inward and then, starting with the bottom side of the leaf, start to roll the leaf upward to encase the filling, tightly tucking the leaf underneath as you move. Transfer to the baking sheet seam side down and prepare the remaining cabbage leaf rolls.

In a deep, large saucepan, warm **2 Tbsp extra-virgin olive oil** over medium-high heat. When the oil is hot, add **1 large yellow or white onion, diced**, and sauté until translucent, 4 to 5 minutes. Add **2 garlic cloves, grated**; **1 Tbsp peeled and grated ginger**; **1 tsp Kashmiri chilli powder (or ¾ tsp ground smoked sweet paprika + ¼ tsp ground cayenne)**; and **½ tsp ground turmeric**. Sauté until fragrant, 30 to 45 seconds. Scrape the bottom of the pot and stir in **one 28 oz [794 g] can diced tomatoes, preferably San Marzano**. Bring to a boil, then lower the heat to a simmer. Taste and season with **fine sea salt** and **a pinch of sugar**, if desired.

Gently and very carefully nestle the stuffed cabbage rolls about halfway in the sauce. Because saucepans are round, use the shape of the pan to guide you. It's OK if it's a tight fit. Cover the saucepan with a lid and simmer until the cabbage is very tender and soft, 25 to 30 minutes. Remove from the heat.

Prepare the tadka. In a small saucepan, warm **1 Tbsp extra-virgin olive oil** over medium-high heat. When the oil is hot, add **1 tsp whole cumin seeds** and **1 tsp whole nigella seeds**. Fry until fragrant and lightly brown, 30 to 45 seconds. Remove from the heat and quickly drop in **½ tsp red pepper flakes such Aleppo, Maras, or Urfa**. Swirl the saucepan until the oil turns slightly red, 15 to 20 seconds. Quickly pour the hot oil over the cabbage rolls. Serve hot or warm. Store leftovers in an airtight container for up to 4 days in the refrigerator.

This is the kind of food that I enjoy making with family and friends who love to cook. I set out small stations for filling and rolling the cabbage rolls and then cook them together in a large pot. And when I need to do this by myself, it is still one of the most relaxing ways to pull my attention away from the everyday hustle and bustle of life.

If you end up with extra filling for whatever reason (maybe you got tired of making the rolls; I've been there and done that), then make the cigar-shaped logs and bread them using the same technique and breading mix used for the Golden Za'atar Onion Rings (page 36) and present them as vegetarian croquettes to your dinner guests.

THE COOK'S NOTES

- Everyone's ratios of salt to water vary when boiling food. I use 1 tsp fine sea salt for every 4½ cups [1 L] water.

- The size of the leaves will start to get smaller as you work down to the core. Consequently, the number of rolls you end up with will vary a little. This recipe gives you enough wiggle room.

- The 1 in [2.5 cm] V-shape cut or notch made at the bottom of the cabbage leaf is important. I've provided a measurement to give you a sense of how small it needs to be. That tough white stalk is like a very tough rubber band and if left attached, it will keep unraveling the roll.

Pasta with Broccoli Miso Sauce

MAKES 4 SERVINGS

Bring a large pot of salted water to a boil over high heat. Add **1 lb [455 g] dried rigatoni or spaghetti** and cook until al dente, per the package directions. Reserve 1 cup [240 ml] pasta water and drain.

Meanwhile, bring another large saucepan or Dutch oven filled with salted water to a rolling boil over high heat. Add **1 lb [455 g] broccoli florets, cut into bite-size pieces**. Boil until tender, 3 to 5 minutes. Transfer to a medium bowl with a slotted spoon and discard the cooking water.

In the same saucepan you used to cook the broccoli, warm ¼ **cup [60 ml] extra-virgin olive oil** over medium heat. When the oil is hot, add **2 Tbsp white or yellow miso paste**; **2 garlic cloves, grated**; **1 Tbsp coarsely ground black pepper**; and **1 tsp red pepper flakes such as Aleppo, Maras, or Urfa**. Sauté until fragrant, 30 to 45 seconds. Fold in the cooked broccoli. Taste and season with **fine sea salt**.

Quickly fold in the hot cooked rigatoni and **1 cup [60 g] grated Parmesan**. Add ¼ cup [60 ml] reserved pasta water, 1 Tbsp more at a time as needed, and stir to create a glossy coating. A large portion of the broccoli will fall apart to form the sauce. Garnish with **2 Tbsp chopped preserved lemon peel, rinsed and drained**. Serve immediately. Leftovers can be stored in an airtight container in the refrigerator for up to 3 days.

While tomatoes reign supreme when it comes to vegetables that make pasta sauce, broccoli is equally spectacular. Broccoli florets are first softened in water by boiling and then cooked with miso, garlic, and Parmesan to form a sauce that hugs the pasta. This intensely savory sauce is inspired by the legendary Italian food writer Marcella Hazan's sugo di broccoli e acciughe from her book, *The Classic Italian Cookbook*.

THE COOK'S NOTES

- Miso does what anchovies do. They make the sauce sing with deep notes of umami.

Cauliflower Bolognese

MAKES 4 SERVINGS

In a large saucepan, warm **2 Tbsp extra-virgin olive oil** over high heat. When the oil is hot, add **1 lb [455 g] coarsely grated cauliflower**. Sauté until lightly browned, 4 to 5 minutes. Remove from the heat and transfer to a bowl.

Wipe the saucepan clean and warm **2 Tbsp extra-virgin olive oil** over medium heat. Add **¼ cup [35 g] finely diced white or yellow onion**; **1 small carrot, finely diced**; and **1 medium stalk celery [30 g], finely diced**. Sauté until tender, 5 to 6 minutes, stirring often to prevent scorching. Stir in **2 Tbsp white miso**, **1 Tbsp low-sodium soy sauce**, and **⅛ tsp ground nutmeg**. Cook until completely combined, with no lumps, 45 seconds to 1 minute. Stir in **one 28 oz [794 g] can crushed tomatoes, such as San Marzano**; **½ cup [120 ml] milk, dairy or plant-based such as oat**; and the cooked cauliflower. Turn down the heat to low, cover, and let simmer for 1 hour, stirring occasionally to prevent burning. Taste and season with **fine sea salt**.

About 20 minutes before the sauce is fully cooked, prepare the pasta. Fill a large saucepan with enough water for the pasta. Stir in **1 tsp fine sea salt**. Bring the water to a rolling boil over high heat. Add **1 lb [455 g] fresh tagliatelle or pappardelle** and cook until al dente, per the package instructions.

Transfer the pasta to the Bolognese sauce with a pair of kitchen tongs, reserving 1 cup [240 ml] cooking water. Add **½ cup [30 g] finely grated Parmesan** (use a Microplane). Fold to coat well. If needed, stir in ¼ cup [60 ml] reserved cooking water to thin out the sauce.

Transfer to a large serving dish and garnish with **2 to 3 Tbsp grated Parmesan** (use a Microplane). Serve warm. Store leftovers in an airtight container in the refrigerator for up to 4 days.

It is impossible to not fall in love with pasta alla Bolognese, the classic dreamy ragù made from slowly simmered tomatoes and milk served with pasta. This version strays away from the classic dish but in a good way. Grated cauliflower replaces the meat, and miso and soy are new additions to help amplify the umami in the sauce.

THE COOK'S NOTES

- In a classic meat-based ragù, the meat is first cooked in milk and then in the tomatoes. Because we're using cauliflower here, those two steps can be combined.

Chicken Katsu with Poppy Seed Coleslaw

MAKES 4 SERVINGS

Prepare the coleslaw by combining in a large bowl **1 lb [455 g] finely shredded green cabbage**; **1 medium red bell pepper, cored and cut lengthwise into thin strips**; **1 large (7¾ oz [220 g]) Granny Smith apple, peeled, cored, grated, and squeezed to remove excess liquid**; **6 scallions, both white and green parts, thinly sliced**; and **¼ cup [3 g] loosely packed chopped cilantro**.

In a small bowl, whisk together **½ cup [120 ml] rice wine vinegar**; **2 Tbsp honey or maple syrup**; **1 tsp black poppy seeds**; **½ tsp red pepper flakes such as Aleppo, Maras, or Urfa**; and **½ tsp ground black pepper**. Pour the mixture over the vegetables in the bowl and toss to coat well. Taste and season with **fine sea salt**. Cover with a lid and refrigerate the coleslaw for at least 30 minutes before serving.

To prepare the chicken katsu, in a wide, shallow bowl, whisk together **2 large eggs** and **½ tsp fine sea salt**.

In a large, wide bowl, whisk together **2 cups [120 g] panko, 1 Tbsp onion powder, ½ Tbsp garlic powder, 1 tsp fine sea salt, 1 tsp ground turmeric, ½ tsp ground black pepper**, and **½ tsp ground cayenne**.

Pat dry with clean kitchen paper towels **4 (total weight 1½ lb [680 g]) boneless, skinless chicken breasts**. Put the chicken on a cutting board and cover with a sheet of cling film. Gently pound the chicken breasts with a rolling pin or mallet to flatten them until they are ½ in [13 mm] thick. Remove and discard the cling film. Season each chicken breast lightly on both sides with **fine sea salt**.

Using a pair of kitchen tongs or two forks, dip the pounded chicken in the egg. Tap the chicken on the sides of the bowl to get rid of any excess liquid and then dip the chicken in the panko mixture, and flip to coat well. Tap the chicken to get rid of any excess crumbs and place on a plate.

Place a wire rack on a baking sheet or line a large plate with absorbent paper towels.

In a large, dry cast-iron or stainless-steel skillet, warm **¼ cup [60 ml] neutral oil with a high smoke point such as grapeseed** over medium-high heat. Fry the panko-coated chicken pieces in the hot oil, cooking in batches as needed, until each side turns golden brown, 3 to 4 minutes per side, and the internal temperature reaches 165°F [75°C]. Lower the heat as necessary to keep from scorching and add more oil between batches as needed. Set the fried chicken on the wire rack to drain any excess oil. The chicken can be kept warm in an oven at 200°F [95°C].

Serve the chicken hot or warm with the coleslaw. Leftover chicken and coleslaw go great in sandwiches, wraps, and salads, and for breakfast with the addition of hard-boiled eggs.

Here's a quick and satisfying meal of chicken cutlets breaded to ethereal crispiness with panko bread crumbs. The coleslaw is light and refreshing, with spots of bold and tart sweetness from the apple, while the turmeric adds a splash of sunshine to the chicken.

THE COOK'S NOTES

- You can use a food processor to grate the apple but be sure to peel the apples or the rougher peel will clump and make a mess in the processor.

- If the apple starts to brown, add 1 to 2 Tbsp of the rice vinegar.

8. Cabbage, Bok Choy, Broccoli, Brussels Sprouts, Collards, Cauliflower, Romanesco, Radishes, Arugula, Kale, Mustard Greens + Watercress

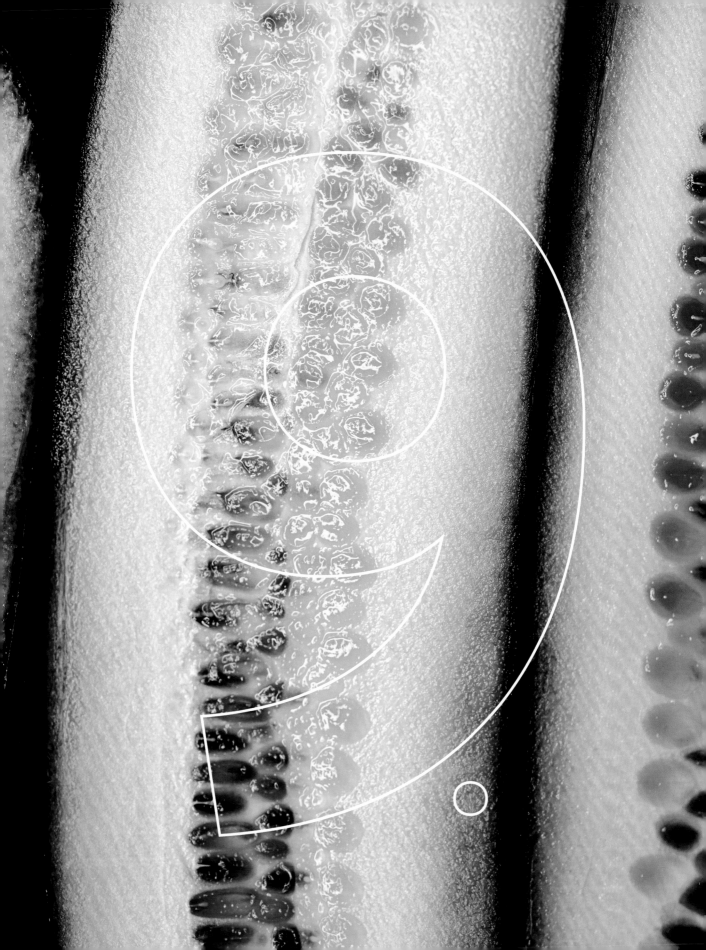

Cucumber Pumpkin Squash + Chayote

The Gourd Family
CUCURBITACEAE

Technically, all the vegetables in this chapter are gourds. However, from a colloquial standpoint, the word *gourd* is used to refer to ornamental squashes that are rarely eaten and primarily reserved for decoration during Halloween and Thanksgiving. Their skin and flesh are much harder and not worth consuming.

Origins
CUCUMBER HAILS FROM INDIA, PUMPKIN ORIGINATED IN MEXICO, AND SQUASH AND CHAYOTE ARE FROM CENTRAL AMERICA.

Cucumber
Buy firm cucumbers, free from damage, but don't press them with force at the market or you will damage them. There are several cucumber varieties to choose from, including English, American, Persian, and lemon. American cucumbers tend to contain softer, edible seeds that don't need to be removed, unlike other varieties.

Squash and Pumpkin
There are two main types of squash—summer and winter. Summer squash are harvested in their immature state before their skin gets hard; winter squash are harvested after ripening when the skin gets hard. Zucchini, chayote, yellow, and pattypan squash are summer squash; pumpkin, butternut, delicata, spaghetti, and acorn squash are winter squash. Winter squash also tend to be rich in carotenes; consequently, their flesh is bright orange-yellow. Outside North America, the word *pumpkin* is also used to describe any type of winter squash.

Avoid squash that isn't firm to the touch; softer spots usually indicate damage. Avoid purchasing summer squash with damaged, slimy, or spotty skin.

The chayote or mirliton reminds me of a large pear with a cucumber-like aroma. There are two types of chayote: spiny/prickly and smooth-skinned. In terms of flavor, I've not noticed a difference; fortunately, the smooth-skinned ones are more common where I live.

Storage
Cucumbers from grocery markets are usually sold coated with a thin layer of wax that will keep them fresh for up to a week in the refrigerator's crisper drawer. Rinse them with warm water before use. Unwaxed cucumbers should be used within a couple of days and kept wrapped with cling film. Do not keep cucumbers very cold; this tends to dry them out. Store summer squash in the refrigerator in a loosely closed bag for a couple of days. Winter squash can be stored in a cool dark spot of the pantry for several months.

Cooking Tips

- Some varieties of cucumbers produce a bitter-tasting class of plant steroids called *cucurbitacins*, but most plant breeders selected this trait out; older heirloom varieties carry the bitter trait. If your cucumber tastes bitter, I recommend a tip from my dad. He trims both ends of the cucumber, sprinkles a little salt on the cut sides, and then rubs the cut end with the trimmed part until the cucumber starts to produce foam. The foam is rinsed off, and the cucumber then supposedly tastes less bitter. Another approach is to salt the cucumber just like an eggplant (see Cooking Tips, page 264); the salt will mask the bitter taste.

- Sliced cucumbers can release a lot of water if left to sit in a salad or sandwich for too long. To avoid this, add the cucumber just when ready to serve. Sliced or diced cucumber can be tossed with salt and left to stand for 10 minutes to remove some liquid via osmosis. Rinse and pat dry before using. Don't overdo the salting time or the cucumber will turn very limp. In sandwiches, a layer of fat—butter, cream cheese, cheese—or fresh lettuce will help block any water from making the bread soggy by acting like a raincoat.

- Use pickling cucumbers for pickling in vinegar, sugar, and spices. They are bred to be firmer and sold immature, which prevents them from falling apart under the harsh conditions of pickling. If treated properly, they even maintain their crunch.

- Cucumber seeds can be hard to chew or introduce too much water to a dish. To remove the seeds from cucumbers, squash, and pumpkins, use a spoon. For cucumbers, slice the cucumber in half lengthwise, then run a teaspoon along the seedy center to scoop out the seeds. For squash and pumpkins, remove the seeds using a sturdy tablespoon or serving spoon.

- Cucumbers with thicker skins can be peeled. One option is to peel the skin in strips: Use a channel knife or citrus zester, or run the prongs of a fork along the length to create a striped pattern.

- Pumpkin seeds are edible and tasty. Pepitas are a type of pumpkin seed that comes from a specific variety of pumpkin called the Styrian or oil-seed pumpkin. The hull of the seed is also edible, and some folks love to eat them with the hull on. To roast pumpkin seeds, first brine them in salted water and then roast them (see below). I remove the strings (add them to the vegetable stock, page 337), as they will burn and turn bitter.

- While the skin of all squash types is edible, I remove the skin only if I feel it is too hard to be eaten comfortably. Winter squashes are characterized by thicker skins, but some squashes, such as delicata, have skins thin enough to be eaten when cooked.

- Summer squashes don't need to be peeled, as their skin is very thin. Sauté in olive oil, butter, or your favorite fat until tender before eating.

- Winter squashes contain more sugar than summer squashes. Roasting winter squashes prior to making soup helps intensify their sweetness and improves their texture (see Yellow Curry Pumpkin Rice Soup, page 200).

- Because winter squashes are rich in pectin and sugar, adding a pinch of baking soda to the vegetable during roasting or pressure cooking helps soften the pectin and accelerates caramelization and the Maillard reaction, producing a range of splendid bittersweet flavors.

- In most recipes that call for pumpkin, other varieties like butternut or acorn can be substituted with ease.

- "Noodles" or "pasta" made from zucchini or yellow squash can work well in recipes. Use a vegetable spiralizer to produce the noodles from the raw vegetable, then toss them into your sauce. Keep in mind, once they're mixed with the sauce, they need to be eaten quickly or they will turn soggy. Try squash noodles in recipes like the Cauliflower Bolognese (page 180) or the Pasta with Broccoli Miso Sauce (page 179). You can also cut the squash into thin lengthwise sections and use them instead of pasta in the Lentil Lasagna (page 242) and bake the dish for a significantly shorter period—I find half the cooking time works.

- Regardless of whether you use spiny or smooth chayote, peel and discard the skin and the inner pit before cooking.

- The sap from chayote contains a type of protein-digesting enzyme called *protease* that also acts as a skin irritant. Use dish gloves and peel the chayote under running water to avoid physical contact with the enzyme. Heating destroys the enzyme and renders chayote safe to eat (see Chayote Chicken Soup, page 207).

Roasted Pumpkin Seeds

This master recipe for Roasted Pumpkin Seeds is fun to do with kids after you have carved a pumpkin for Halloween.

YIELD VARIES

Preheat the oven to 375°F [190°C]. In a small saucepan, bring **2 cups [480 ml] water** and **1 Tbsp fine sea salt** to a boil over medium-high heat. Add **1 cup [140 g] raw, unhulled pumpkin seeds (strings removed and discarded)** and boil for 10 minutes. Drain the seeds in a fine mesh sieve and discard the water. Spread the seeds on a rimmed baking sheet. Drizzle with **1 Tbsp extra-virgin olive oil** and toss to coat. Roast until the seeds turn golden brown and crisp, rotating the pan halfway through the cooking time, 25 to 30 minutes. Watch the seeds carefully after the first 8 to 10 minutes; some of the smaller seeds might finish sooner and will need to be removed. Toss the hot seeds with **1 Tbsp of your favorite spice blend** and **fine sea salt** if needed. Let the seeds cool to room temperature and transfer to an airtight container. Store for up to 1 month in the refrigerator or 3 months in the freezer.

Cucumber + Roasted Peanut Salad

MAKES 4 SERVINGS

In a large mixing bowl, place **2 large English cucumbers, cored and diced**. Add ¼ cup **[35 g] roasted, salted peanuts, crushed**; **1 shallot, thinly sliced**; **1 fresh chilli such as jalapeño or serrano, minced**; and **¼ cup [3 g] packed cilantro leaves**.

In a small bowl, combine **2 Tbsp Chinese black or malt vinegar**, **1 Tbsp sesame oil (preferably toasted)**, **1 Tbsp peeled and grated fresh ginger**, and **½ tsp ground black pepper**.

Drizzle over the cucumber. Toss to coat. Season with **flaky sea salt**. Serve immediately. Leftovers do not store well.

This is a joyful, crunchy salad, a nod to a cucumber salad from the coastal state of Maharashtra in India. Cucumbers in summer are a no-brainer—but here it's the roasted peanuts that make this salad so alluring. I use store-bought roasted peanuts; in summer I want to avoid using the stove.

THE COOK'S NOTES

- Do not allow the finished salad to sit for more than 20 minutes before you eat it. The cucumber will release a lot of water that will make the salad soggy, and the peanuts will turn soft.

- This is one recipe in which malt vinegar works as well as the Chinese black vinegar I call for.

- If you prefer this less hot, use a milder chilli or half a chilli, or deseed before use.

Grilled Zucchini with Pumpkin Seed Chutney

MAKES 4 SERVINGS

Pumpkin Seed Chutney

In a dry stainless-steel skillet over medium heat, dry-roast **2 cups [280 g] raw unsalted pumpkin seeds** until the seeds just start to turn brown, about 2 minutes, then add **1 tsp whole cumin seeds**. Cook until fragrant, about 2 minutes more. Remove from the heat and transfer to a blender along with **1¼ cups [300 ml] boiling water, plus more as needed**; **3 Tbsp fresh lemon or lime juice**; **2 garlic cloves**; and **1 tsp ground chipotle**. Blend until smooth, adding 1 Tbsp water at a time if needed to get things moving. Taste and season with **fine sea salt** and transfer to a serving bowl. Drizzle **2 Tbsp extra-virgin olive oil** over the chutney. The chutney can be made a day in advance. Store in an airtight container for up to 2 days and bring to room temperature before serving.

Heat the grill to high and brush the grates with **a little extra-virgin olive oil**. Brush **4 spring onions, trimmed, both white and green parts, cut in half lengthwise** (see the Cook's Notes), and **8 medium zucchini, cut lengthwise into ¼ in [6 mm] planks**, with **3 to 4 Tbsp [45 to 60 ml] extra-virgin olive oil**. Season with **fine sea salt** and **freshly ground black pepper**. Grill the vegetables, turning occasionally, until they develop dark char marks and become tender, about 3 minutes per side. The onions' color will become more vibrant, while the zucchini flesh will turn pale yellow. Transfer the vegetables to a serving platter.

Serve the chutney alongside the grilled vegetables.

Zucchini, a summer squash, was made for grills and barbecue parties. It doesn't need much more than oil and seasoning, and it cooks quickly. Fresh spring onions are also seared on the hot grates of the grill, and both vegetables are served with a pumpkin seed chutney that's got a spark of chipotle in it. This chutney goes great with any kind of grilled vegetable. Swap out the spring onions and zucchini with other in-season vegetables, such as eggplant in summer—and yes, it even goes well with wedges of grilled winter pumpkin.

THE COOK'S NOTES

- The name *spring onion* causes a lot of confusion. What I'm referring to here are young onions with enlarged bulbs and green stalks still attached—not scallions.

- The cooking time will vary depending on the size of your onions. Larger, thicker bulbs take longer to cook.

- If you don't own a grill, use a grill pan over medium-high heat.

190

Acorn Squash, Kale + Chilli Miso Sauce

MAKES 4 SERVINGS

Preheat the oven to 400°F [200°C].

Prepare the chilli miso sauce. In a large stainless-steel saucepan, warm **¼ cup [60 ml] neutral oil with a high smoke point such as grapeseed** over medium-high heat. When the oil is hot, add **1 large red onion or 6 shallots, finely chopped**, and **4 garlic cloves, minced**, and sauté until they start to turn golden brown, 6 to 8 minutes. Stir in **1 Tbsp white or yellow miso paste, 2 tsp smoked sweet paprika, 1 tsp hot red pepper flakes** (see the Cook's Notes), and **1 tsp ground cumin** and cook until the oil starts to turn bright red and the mixture bubbles, 1 to 2 minutes. Remove from the heat and transfer to a small bowl. Stir in **2 Tbsp rice or Chinese black vinegar; 1 Tbsp low-sodium soy sauce;** and **2 tsp maple syrup, date syrup, or honey**. The sauce can be made a week in advance (see the Cook's Notes).

On a rimmed baking sheet or roasting pan, toss **2 lb [910 g] acorn squash, cut in half crosswise, seeds removed, and sliced into 1 in [2.5 cm] half rings**, with **1 Tbsp neutral oil with a high smoke point such as grapeseed** and **fine sea salt**. Roast for 30 to 45 minutes, until golden brown, flipping halfway through cooking. Remove from the oven.

While the squash roasts, prepare the kale. In a 12 in [30.5 cm] cast-iron or stainless-steel skillet, warm **1 Tbsp neutral oil with a high smoke point such as grapeseed** over medium-high heat. When the oil is hot, add **1 bunch dinosaur kale (10½ oz [300 g]), finely shredded**, and **fine sea salt**. Use tongs to toss and sauté until the leaves turn brilliant green and the stems are tender, 3 to 5 minutes. Remove from the heat. Drizzle with **1 Tbsp rice vinegar or Chinese black vinegar**. Taste and season with **fine sea salt**.

To serve, transfer the kale to a serving plate. Place the cooked squash slices on top. Top with half of the chilli miso sauce and serve the rest on the side. Garnish with **2 scallions, both white and green parts, thinly sliced; ¼ cup [3 g] packed cilantro leaves;** and **¼ cup [35 g] crushed roasted unsalted peanuts** and serve immediately.

continued

I used to avoid winter squashes because I found them too sweet. But after years of staying away, I realized the problem. Every time I tried them in restaurants, they were dressed in a cloyingly sweet sauce that could have been passed off as dessert. My palate reached peak sweet + squash fatigue. It was only after I started cooking squash on my own that I really began to understand what worked for me. This acorn squash is a good example of why I've changed my ways. The chilli miso sauce slathered over roasted crescents of acorn squash with hints of vinegar is a perfect combination, balancing the sweetness of the squash itself.

THE COOK'S NOTES

- The chilli miso sauce can be made at least a week in advance and stored in an airtight container in the refrigerator. Bring to room temperature before use. It also goes well over grilled cabbage, eggs, potatoes, and sandwiches.

- I prefer using white or yellow miso paste because they pack a deep umami punch while contributing less salt than red miso and give me more control over the salting of the dish.

- While I normally recommend using Aleppo, Maras, or Urfa in place of red pepper flakes, this is one recipe where I feel the sauce benefits from using regular hot red pepper flakes—like the kind that comes with your pizza.

- Besides acorn squash, do consider using delicata, pumpkin, or butternut here.

194

Butternut Squash Sauce, Crispy Leeks + Farfalle

MAKES 4 SERVINGS

Preheat the oven to 400°F [200°C].

On a rimmed baking sheet, toss together **1 lb [455 g] peeled and seeded butternut squash, diced into 1 in [2.5 cm] cubes**, and **1 Tbsp extra-virgin olive oil**. Spread in a single layer and roast until completely tender and lightly caramelized, 30 to 45 minutes. Remove from the oven and transfer to a medium saucepan.

To the saucepan, add **2 cups [480 ml] water**; **½ cup [70 g] unsalted raw cashews**; **2 garlic cloves, smashed**; **½ tsp ground turmeric**; **¼ tsp ground cayenne**; **¼ tsp baking soda**; and **⅛ tsp ground nutmeg**. Bring to a boil over medium-high heat. Remove from the heat and let sit for 5 minutes.

Transfer the mixture to a high-speed blender or food processor. Add **3 Tbsp white or yellow miso** and, using caution, carefully blend the hot liquid until silky smooth. Keep the top of the blender covered with a dish towel to help keep the sauce warm.

While the squash and cashew mixture is simmering, bring a large pot of salted water to a boil and cook **1 lb [455 g] farfalle** to al dente, per the package instructions. Reserve 1 cup [240 ml] of the pasta water, then drain the cooked pasta and transfer to a large bowl.

Meanwhile, in a medium saucepan, warm **¼ cup [60 ml] extra-virgin olive oil** over medium-high heat. Lower the heat to medium-low and add **2 medium leeks, trimmed and thinly sliced**, and **¼ tsp fine sea salt**. Sauté until lightly golden brown and starting to crisp, 10 to 20 minutes (this might take longer and, just like browning onions and life, is unpredictable when it comes to time). If they start to burn, add 1 Tbsp water and scrape the bottom of the pot. Add **2 Tbsp raw unsalted pumpkin seeds** and **1 tsp red pepper flakes such as Aleppo, Maras, or Urfa**. Sauté until the seeds start to brown, 1 to 2 minutes. Remove from the heat, taste, and season with **fine sea salt**.

When ready to serve, fold the pasta into the sauce, adding some of the reserved pasta water, 1 to 2 Tbsp at a time, to make the sauce glossy (the sauce will become drier as the pasta sits). Top with the crispy leeks and pumpkin seeds. Garnish with **4 large fresh sage leaves, thinly sliced**.

continued

Serve hot or warm. Store leftover pasta in an airtight container in the refrigerator for up to 3 days. When reheating, add 1 to 2 Tbsp water to rehydrate the pasta.

Farfalle is my favorite pasta shape, quite possibly because it's one of the most playful ones. So there was no way I'd leave it out of this book. This pasta sauce is remarkably easy to put together and produces superior results. But, best of all, it takes less than an hour to prepare, making it a great meal for a busy weekday. Roast, blend, and season the butternut squash to make the sauce, and top the pasta with plenty of crispy leeks.

THE COOK'S NOTES

- What holds this pasta sauce together is the combination of butternut squash, cashews, and garlic. They each contain emulsifying agents that make the sauce cling to the noodles.

- Baking soda and salt help soften the tough pectin in the squash and pave the way for creating a velvety pasta sauce.

- A high-speed blender does a fantastic job of making a smooth purée. If you don't own one or your blender isn't the strongest, substitute ½ cup smooth cashew butter in place of the cashews.

- Shallots are an excellent alternative to leeks here.

- For a "cheesy" taste, blend in 1 Tbsp nutritional yeast with the roasted squash.

Chilled Cucumber Soup with Jalapeño Oil

MAKES 2 SERVINGS AS A MEAL AND 4 SERVINGS AS A STARTER

Make a batch of jalapeño oil (see Creamy Corn Soup with Jalapeño Oil, page 77).

Cut in half crosswise **1 large English cucumber**. Dice one half and reserve in a small bowl. Cover and refrigerate for garnish. Peel the other half. Set a fine mesh sieve over a serving bowl. In a blender, combine the peeled cucumber with **3 cups [720 g] chilled plain, unsweetened full-fat yogurt**; **½ cup [70 g] whole raw unsalted cashews**; **4 garlic cloves**; **2 Tbsp fresh dill**; **1 Tbsp extra-virgin olive oil**; **2 tsp fresh lemon juice**; and **½ tsp ground black pepper**. Blend on high speed until smooth. Taste and season with **fine sea salt**. Strain the soup through the fine mesh sieve. Cover and refrigerate for at least 1 hour and up to 4 hours.

When ready to serve, stir the soup well and divide between two or four bowls. Garnish each bowl with the reserved diced cucumber and about ½ **tsp fresh dill leaves**. Drizzle with **1 to 3 tsp jalapeño oil** and serve immediately.

Since moving to Los Angeles, I've come to appreciate the brilliance of chilled soups. Our summers are hot and there are days when the only thing I want to touch is my freezer—or better still, a bowl of this chilled cucumber soup. Cashews and garlic help create a creamy, velvety emulsion of the cucumber while the dill raises the refreshing factor. I like to serve this as a starter to a large meal or on its own (serving sizes change accordingly).

THE COOK'S NOTES

- This soup should be consumed only cold or chilled; it will not taste as pleasant and refreshing if warmed to room temperature or higher. Chilling all the ingredients and freezing the cashews prior to blending help.

Yellow Curry Pumpkin Rice Soup

MAKES 8 CUPS [1.9 L]

Preheat the oven to 400°F [200°C]. Line a baking sheet or roasting pan with foil.

Cut a **2 lb [910 g] sugar pie pumpkin or kabocha squash** in half crosswise and remove the seeds (see Roasted Pumpkin Seeds, page 187). Place on the prepared baking sheet and brush the cut side of the pumpkin with **1 Tbsp extra-virgin olive oil** and sprinkle with **1 tsp fine sea salt**. Roast until the pumpkin is tender, 35 to 45 minutes, rotating the sheet halfway through. The flesh should turn golden brown and be easily pierced by a knife or skewer. Remove from the oven and let rest until the pumpkin is cool enough to handle. Separate the flesh from the skin.

While the pumpkin roasts, in a medium heatproof bowl, steep **10 to 12 dried shiitake mushrooms (about 1 oz [30 g])** in 2 cups [480 ml] hot water for 30 minutes. The liquid will turn dark brown. Squeeze the mushrooms to extract as much liquid as possible and discard the mushrooms. Reserve the steeped liquid.

In a medium saucepan, warm **2 Tbsp extra-virgin olive oil** over medium-high heat. When the oil is hot, add **2 Tbsp yellow curry paste** and **1 garlic clove, chopped**. Sauté, stirring constantly, until the paste is fragrant and starts to stick to the bottom of the saucepan, 1 to 1½ minutes. Stir in the steeped mushroom liquid (be careful of splattering) and the cooked pumpkin. Remove from the heat.

Using an immersion blender, pulse the mixture until smooth and velvety. You should have about 3 cups [710 ml]. Add enough water to bring the final volume up to 7 cups [1.7 L]. The soup can also be prepared using a blender or food processor, working in batches as needed.

Return the blended soup to the stove over medium-high heat. Stir in **2 Tbsp peeled and grated fresh ginger** and **1 Tbsp fresh lime juice**. Fold in **1 cup [150 g] cooked black or forbidden rice (see page 205)**. Taste and season with **fine sea salt**.

Serve hot or warm, topping each bowl with a **small handful of microgreens such as arugula or radish**. Leftover soup can be stored in an airtight container in the refrigerator for up to 3 days.

continued

I like soups that are hearty with plenty of texture, but also full of vibrant colors and flavor. The turmeric in the yellow curry paste, the bright orange flesh of winter squash, the dark purple grains of forbidden rice, and the bright green of microgreens all make this a beautiful sight to behold. A big spoon to eat this up with is all that you need.

- There are lots of great options for Thai curry pastes, but my recommendation is Mekhala. It is a stellar brand to keep in your pantry.

- Carrots, potatoes, and sweet potatoes can also be used in this soup.

- Forbidden rice adds a lovely pop of color, but you can use white or brown rice too.

- Steeping the shiitake mushrooms in hot water helps produce a tea full of umami goodness. Among mushrooms, dried shiitakes contain the highest quantity of glutamates, so they are my top choice. If you want to be even more precise, use the method of steeping listed for the Master Mushroom Vegetable Stock (page 337), which relies on using exact temperatures for maximum and efficient glutamate production.

202

Chana Masala Pumpkin Pots

MAKES 4 SERVINGS

To prepare the chana masala, in a large saucepan or Dutch oven, warm **2 Tbsp extra-virgin olive oil or ghee** over medium-high heat. Add **2 large yellow or white onions, thinly sliced**, and sauté until they turn translucent, 4 to 7 minutes. Add **2 garlic cloves, minced**; **1 Tbsp peeled and grated fresh ginger**; **1 tsp garam masala, homemade (page 341) or store-bought**; **1 tsp whole cumin seeds**; and **½ tsp ground turmeric** and cook until fragrant, 30 to 45 seconds. Add **¼ cup [60 g] tomato paste** and cook until the tomato paste starts to brown, 2 to 3 minutes, lowering the heat as necessary to avoid scorching. Fold in **two 14 oz [400 g] cans chickpeas, rinsed and drained**, and **1 cup [240 ml] water** and heat until the chickpeas are warmed through.

Using an immersion blender, pulse for a few seconds to break up some of the chickpeas. This will help thicken the liquid. Don't overdo this, but if you end up making it too thick, stir in ½ cup [120 ml] water. Stir in **1 Tbsp lemon juice** and scrape the bottom of the pot to incorporate any browned bits that remain. Taste and season with **fine sea salt**.

Preheat the oven to 400°F [200°C].

Cut about 1 in [2.5 cm] of the top of **2 small sugar pie pumpkins (each about 2 lb [910 g])** to form a lid. Core and remove the seeds (see Roasted Pumpkin Seeds, page 187) and stringy materials (freeze and use to make vegetable stock, page 337). Brush the inside of the pumpkin and the tops with **2 Tbsp extra-virgin olive oil** and sprinkle with **fine sea salt**. Place the pumpkins and their lids, cut side up, on a rimmed baking sheet or roasting pan and roast until the flesh is tender and easily pierced by a skewer or fork and golden brown, 40 to 50 minutes. Remove from the oven.

While the pumpkins roast, cook the rice. In a medium saucepan, bring **1 cup [100 g] black or forbidden rice, rinsed and drained**; **2½ cups [600 ml] water**; and **fine sea salt** to a boil over medium-high heat. Turn down the heat to low and simmer, covered, until all the water is absorbed and the grains are tender, 40 to 50 minutes. Remove from the heat and let rest, covered, for 15 minutes.

continued

To serve, transfer the rice to a serving plate. Carefully transfer the pumpkins on top of the rice (be careful, they will be very delicate once roasted and can easily collapse; a sturdy slotted fish spatula will be your friend here). Divide the chana masala between the two pumpkins. Garnish the chana masala with a **few sprigs of cilantro** and **2 scallions, both white and green parts, thinly sliced**. Top the rice with **2 Tbsp toasted salted pumpkin seeds**.

Serve immediately with a carving knife and a large serving spoon. Leftovers can be stored in an airtight container in the refrigerator for up to 3 days.

Orange and black! Visually, everything about this dish screams Halloween to me and I even considered carving these little pumpkin pots into jack-o'-lanterns. In this recipe, roasted sugar pie pumpkins are filled with chana masala and then served over a bed of warm forbidden rice. For Thanksgiving or any fall dinner, this makes a delicious and seasonally spectacular entrée.

THE COOK'S NOTES

- If you don't want to serve these as "pots," slice the pumpkin into thick wedges and roast them in the oven. Top the wedges with the chana masala.

- Forbidden rice or black rice is a medium-grain rice that gets its dark purple—almost black—color from the anthocyanin pigment. Buy it online or in Asian markets.

- Tomato paste is an essential staple to keep in your pantry. It's a wonderful shortcut when you need a concentrated spot of tomato flavor and don't have the time to cook down tomatoes for a long period. Buy it in tubes; they're easier to use and store than cans.

Chayote Chicken Soup

MAKES 4 SERVINGS

In a large saucepan or Dutch oven, warm **2 Tbsp extra-virgin olive oil** over medium-high heat. When the oil is hot, add **2 skin-on, bone-in chicken breasts (total weight about 1½ lb [680 g])**, placing them skin side down. Sear until the skin turns golden brown, 3 to 4 minutes. Transfer the chicken to a plate, reserving the fat in the saucepan.

To the hot pan, add **3 or 4 medium carrots, peeled and diced**, and **1 large white or yellow onion, diced**. Sauté until the onions become translucent and the carrots just start to brown, 3 to 4 minutes. Add **2 garlic cloves, minced**; **one 2 in [5 cm] stick of cinnamon**; **1 tsp ground cumin**; **1 tsp ground coriander**; and **¼ tsp ground cayenne**. Sauté until fragrant, 30 to 45 seconds.

Add **2 medium chayote, peeled, cored, and diced into ¼ in [6 mm] cubes**; the chicken; **5 cups [1.2 L] water**; and **1 tsp fine sea salt**. Bring to a boil over high heat, then lower the heat and simmer for 1 hour, skimming the foam occasionally, until the chicken is completely cooked and the chayote tender.

Discard the cinnamon. Transfer the chicken to a plate. Shred the chicken with a pair of forks and return the chicken with all the liquid back to the soup. Discard the bones and skin. Stir in **1 Tbsp fresh lime or lemon juice** and **2 Tbsp chopped cilantro**. Taste and season with **fine sea salt** and additional **lime juice, to taste**.

Serve hot or warm. Store leftovers in an airtight container for up to 3 days in the refrigerator.

This is your new go-to chicken soup when you need a pick-me-up, when you crave something satisfying and delicious, or when you want to sit on the couch wrapped in a blanket, avoiding the snow or rain. There is a pure feeling of satisfaction that emanates from the tender texture of the chayote along with the fragrant spices and hints of fresh lime in this soup. This soup will make you happy.

THE COOK'S NOTES

- For a shortcut, use chicken stock and store-bought or homemade leftover rotisserie chicken.

- Browning the chicken skin is essential for developing the flavors of the soup.

- Wear gloves and peel the chayote under running water. The sap from the vegetable can irritate the skin (see Cooking Tips, page 187).

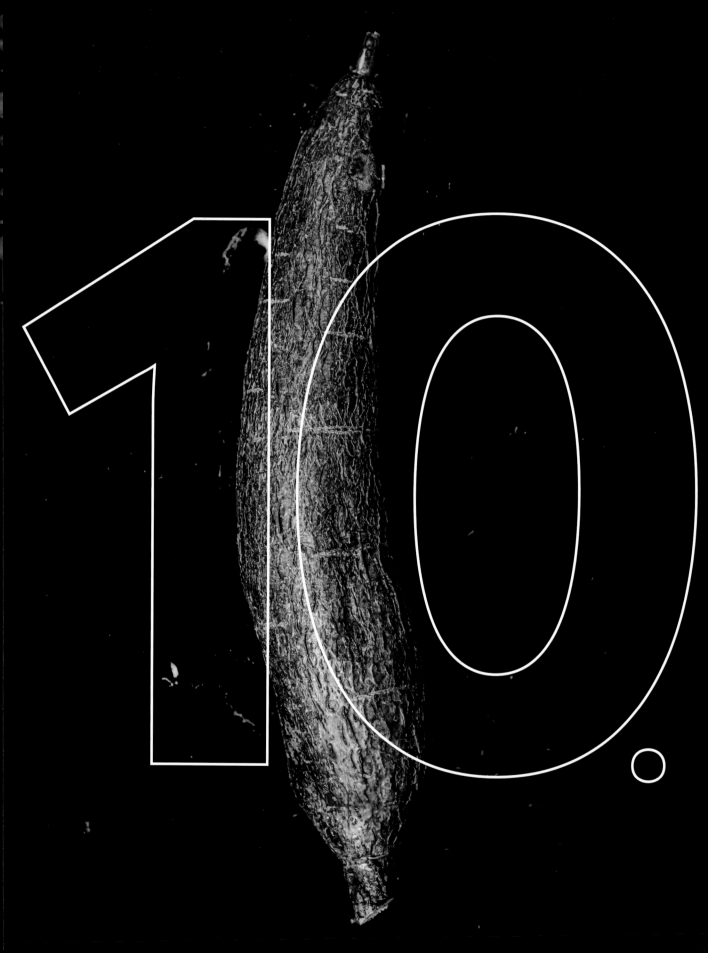

Cassava

The Spurge Family
EUPHORBIACEAE

Origin

CASSAVA ORIGINATED IN SOUTH AMERICA.

Cassava

Cassava, manioc, or yuca is not to be confused with yucca (a different plant that belongs to the asparagus family; see page 84), seen in many Southern California gardens and throughout Mexico. The Swedish biologist Carl Linnaeus, who was responsible for the taxonomy of plants and animals, erroneously muddled the two plants after receiving a flattened dried specimen of the yucca plant. Thinking it was yuca, he christened it with the scientific genus name Yucca, and it stuck. I generally use the word *cassava* throughout this book to avoid any confusion. Cassava is a source of starch and the source for tapioca.

Storage

Cassava is often sold with a thin coat of edible wax to help increase its shelf life. Unpeeled cassava must be stored in a cool, dark spot in the kitchen. Peeled cassava can be stored submerged in fresh water for up to 3 weeks; discard and refresh the water daily.

Cooking Tips

- Cassava must be cooked before eating. It contains linamarin and lotaustralin, naturally occurring forms of cyanide that are toxic if consumed raw and in large quantities.

- This method of preparing cassava works for both fresh and frozen varieties. If starting with fresh, wash the tuber under hot running water to remove any wax. Peel and discard the outer brown skin. Cut in half lengthwise and remove and discard the tough inner string that runs through the center. Cut the cassava into 2 in [5 cm] pieces as needed and place them in a large enough saucepan. It is best to cut larger chunks and boil to avoid excess mushiness; if they need to be smaller for a recipe, do that after boiling. Cover with enough water to submerge them by 1 in [2.5 cm] and add fine sea salt (1 tsp fine sea salt to every 4½ cups [1 L] water). If you plan on roasting the cassava (see Cassava Bravas, page 212), add ⅛ tsp baking soda to the salted water during boiling; this will give a crispier coat. Bring the water to a rolling boil over high heat, then turn down the heat to low and simmer, covered, until the cassava is tender and easily pierced by a fork, 20 to 30 minutes. Remove from the heat, drain the water, and use the cassava as needed. This can be done a day ahead of time; store the drained cassava in an airtight container in the refrigerator.

- If you use frozen cassava, it is best to cut the cassava and remove the tough inner string after it is cooked and tender.

- Cassava is rich in starch and can be used just like potatoes or flour to thicken stews and soups. It will absorb flavors easily, so taste the dish and season as needed.

Cassava Bravas

In a large saucepan, place **2 lb [910 g] peeled cassava, inner string removed, and cut into 2 in [5 cm] chunks**, and add enough water to cover them by 1 in [2.5 cm]. Add **1 tsp fine sea salt, ½ tsp ground turmeric, and ⅛ tsp baking soda**. Bring to a boil over medium-high heat. Lower the heat to a simmer and cook, covered, until the cassava is tender and easily pierced by a fork or knife, 20 to 30 minutes. Remove from the heat, drain, and discard the water. (If frozen cassava was used, when cool enough to handle, remove the inner string.)

Preheat the oven to 400°F [200°C].

Transfer the cassava to a large mixing bowl. Add **2 Tbsp extra-virgin olive oil, 1 tsp ground black pepper**, and **fine sea salt**. Toss well or gently fold with a silicone spatula to coat evenly; some of the cassava will form a thick paste that will coat the pieces and eventually form a crust.

Spread the cassava in a single layer on a rimmed baking sheet and cook, flipping halfway through with a thin spatula or kitchen tongs, until golden brown and crisp all over, 35 to 40 minutes. Remove from the oven and transfer to a serving bowl.

While the cassava cooks in the oven, prepare the tomato sauce. In a medium saucepan, warm **2 Tbsp extra-virgin olive oil** over medium-high heat. When the oil is hot, add **2 garlic cloves, grated**; **1 tsp red pepper flakes such as Aleppo, Maras, or Urfa**; **1 tsp ground smoked sweet paprika**; and **1 tsp whole cumin seeds**. Cook until fragrant and the oil starts to turn red, 30 to 45 seconds. Stir in **one 14 oz [400 g] can crushed tomatoes** and **1 tsp sherry or red wine vinegar**. Bring to a boil, turn down the heat to low, and simmer for 5 minutes. If the sauce is thin or watery, whisk in **1 tsp all-purpose flour or cornstarch** until slightly thickened, just enough to bind all the loose water. If it needs further thickening, whisk in **½ tsp flour**. Taste and season with **fine sea salt**. Remove from the heat.

To prepare the aioli, in a medium bowl, whisk together **1 cup [240 g] mayonnaise**; **4 garlic cloves, grated**; **½ tsp ground smoked sweet paprika**; and **½ tsp ground chipotle**. Taste and season with **fine sea salt**.

212

To serve, top the cassava with 2 to 3 Tbsp of the tomato sauce and 2 to 3 Tbsp of the aioli. Garnish with **2 Tbsp chopped flat-leaf parsley** and serve immediately with the extra sauce and aioli on the side. Store leftovers in the refrigerator in an airtight container for up to 3 days.

This is my rendition of that great Spanish dish, patatas bravas, but with a few differences, a major one being the use of cassava in place of potatoes. Cassava is first boiled and then roasted in the oven until crispy. It's served with two sauces: a rich tomato sauce flavored with vinegar and a smoky aioli. Use good-quality paprika and chipotle that are rich in their smoky profile.

THE COOK'S NOTES

- In this recipe, the cassava is boiled a little differently than in the following two recipes. I'm borrowing a technique from cookbook author Kenji López-Alt that he developed to make crispy potatoes. Using baking soda as well as salt helps create a crispier texture by softening the pectin that, when baked, makes it crunchy and crispy. Don't panic if the water turns blood red; it's the curcumin (the yellow pigment in turmeric) responding to the change to an alkaline pH from the baking soda. The cassava will eventually turn yellow when roasted.

- Adding turmeric to the boiling water stains the cassava with a golden yellow hue.

- The acids in the tomatoes and the vinegar can sometimes decrease the ability of the starch to thicken the tomato sauce. However, it is best to start with a small quantity of flour or cornstarch and then add ½ tsp increments until you achieve the desired consistency, or you could end up with a firm pudding.

- Use whatever mayonnaise tickles your fancy, including your favorite plant-based mayonnaise to veganize this recipe.

- The smoky flavor in the tomato sauce and the aioli comes from ground sweet paprika. Chipotle adds an extra bang of smokiness to the aioli.

Bombay Masala Cassava

MAKES 4 SERVINGS

In a 12 in [30.5 cm] cast-iron or stainless-steel pan, warm **2 Tbsp neutral oil with a high smoke point such as grapeseed** over medium-high heat. Add **1 medium yellow or white onion, thinly sliced**, and sauté until translucent, 3 to 4 minutes. Add **1 tsp black or brown mustard seeds**; **1 tsp red pepper flakes such as Aleppo, Maras, or Urfa**; **1 tsp garam masala, homemade (page 341) or store-bought**; and **½ tsp ground turmeric** and sauté until fragrant, 30 to 45 seconds, lowering the heat if necessary to avoid scorching.

Turn the heat to medium-high if it's not already and fold in **2 lb [910 g] cassava, cut into 1 in [2.5 cm] chunks and boiled (see page 210)**; if using frozen cassava, it can be cut after boiling. Drizzle with additional oil if the pan seems dry. Season with **fine sea salt** and sauté until the cassava starts to turn golden brown, 3 to 4 minutes. Remove from the heat and transfer to a serving bowl.

Drizzle the cassava with **2 Tbsp fresh lemon or lime juice** and garnish with **1 fresh chilli such as jalapeño, serrano, or bird's eye, thinly sliced**; **1 Tbsp chopped cilantro**; and **1 Tbsp torn fresh mint leaves**.

Serve warm. Store leftovers in the refrigerator in an airtight container for up to 3 days.

Every time I read the title of this recipe, I keep seeing "Masala Casanova," which would be a great title if a dish could serenade. But at a closer glance, there is a little romance brewing here. A loving relationship seems to arise when cassava is cooked with fragrant spices and then finished off with a splash of fresh lemon juice and herbs.

THE COOK'S NOTES

- You can use fresh or frozen cassava. If you find it easier, cut the cassava into larger pieces, boil, and when cool enough to handle, cut them smaller. See instructions on page 210.

- For a nutty flavor, consider swapping in ghee for the grapeseed oil.

Swordfish + Crispy Cassava with Chimichurri

MAKES 4 SERVINGS

In a medium bowl, combine **½ cup [120 ml] extra-virgin olive oil**; **¼ cup [60 ml] red wine vinegar**; **½ cup [10 g] tightly packed cilantro, finely chopped**; **½ cup [10 g] tightly packed flat-leaf parsley, finely chopped**; **4 garlic cloves, grated**; **1 or 2 fresh chillies such as jalapeño, serrano, or bird's eye, finely chopped**; **½ tsp dried oregano**; and **fine sea salt**. Let sit for at least 1 hour before using. This can be made at least 1 day in advance, stored in an airtight container in the refrigerator, and served at room temperature.

Preheat the oven to 400°F [200°C]. Line a rimmed baking sheet with foil.

On the baking sheet, toss together **2 lb [910 g] prepared cassava, cut into 2 in [5 cm] long and ½ in [13 mm] wide sticks and boiled (see page 210; if using frozen cassava, cut it after boiling)**; **2 Tbsp extra-virgin olive oil**; **1 tsp ground black pepper**; and **fine sea salt**, taking care not to break the cassava. Spread in a single layer and cook until golden brown, 30 to 35 minutes, flipping the cassava halfway through cooking. Remove from the oven.

About 15 minutes before the cassava is finished baking, prepare the fish. In a large cast-iron or stainless-steel skillet, warm **2 Tbsp extra-virgin olive oil** over medium-high heat. Pat dry **4 (each 6 to 8 oz [170 to 227 g]) swordfish steaks, about 1 in [2.5 cm] thick**, with paper towels, then season both sides with **fine sea salt** and **ground black pepper**. Cook on each side for 3 to 4 minutes, until a golden-brown crust forms and the fish flakes easily. Remove from the heat, transfer to a plate, and let rest for 5 minutes.

Top the fish with the chimichurri and serve with the crispy cassava. Store leftovers in the refrigerator in an airtight container for up to 3 days.

Chimichurri, the vinegary herb sauce from Argentinian and Uruguayan cuisines, is bold and powerful. This recipe takes advantage of that feature by using a single sauce for the two main ingredients, the swordfish steak and the cassava. One sauce to rule them all!

THE COOK'S NOTES

- You can use fresh or frozen cassava. If you find it easier, cut the cassava into larger pieces, boil, and when cool enough to handle, cut them smaller. See instructions on page 210.

- If you want to skip the fish and just eat the cassava with chimichurri, you can do that. But if you want to add some protein here, my top two choices are cooked black lentils or Crispy Tofu (page 166).

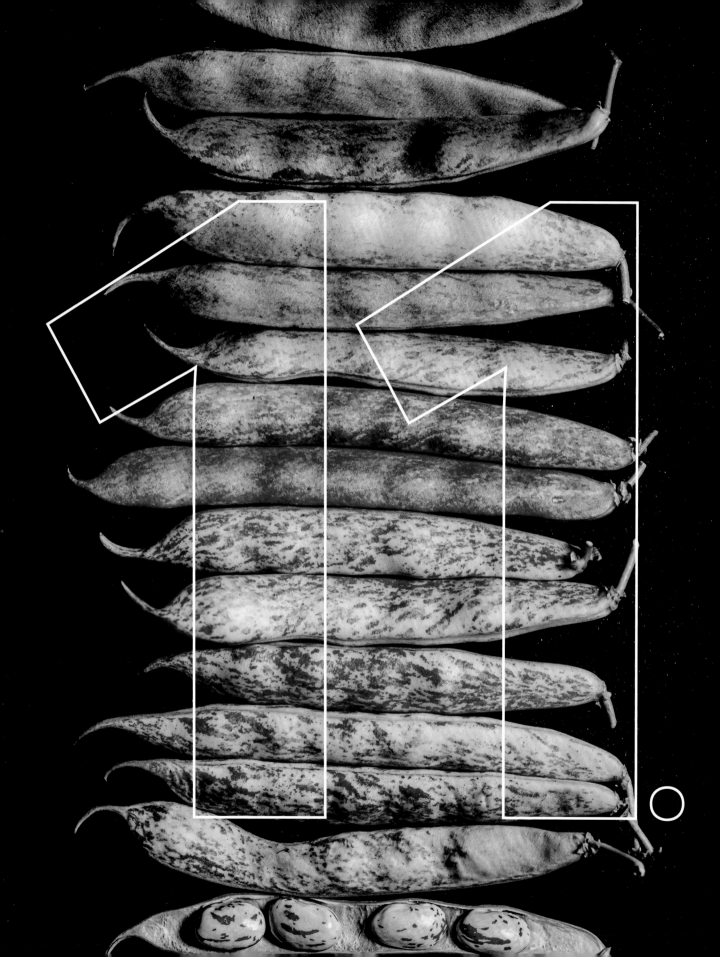

Chickpeas Soybeans Lentils Peas Beans + Jícama

The Pea or Bean Family

FABACEAE OR LEGUMINOSAE

This, the third-largest plant family, includes peas and legumes.

Origins

BEANS OF THE PHASEOLUS FAMILY, INCLUDING BLACK, KIDNEY, PINTO, CANNELLINI, LIMA, AND GREEN BEANS, COME FROM THE AMERICAS. BEANS OF THE VINCA FAMILY, SUCH AS MUNG BEAN, URAD, BLACK-EYED PEAS, AND ADZUKI, COME FROM SOUTH ASIA. PEAS ARE FROM THE MEDITERRANEAN. PIGEON PEAS HAIL FROM INDIA. LENTILS COME FROM WESTERN ASIA AND THE LEVANT. CHICKPEAS ORIGINATED IN TURKEY, THE LEVANT, THE NEAR EAST, AND INDIA. PEANUTS COME FROM SOUTH AMERICA. JÍCAMA ORIGINATED IN MEXICO AND CENTRAL AND SOUTH AMERICAS. SOYBEANS HAIL FROM CHINA.

Fresh and Dried Beans

All beans are created differently, not only in their looks but also in their chemistry. When it comes to cooking, there are two types of beans: easy-to-cook (ETC) and hard-to-cook (HTC). ETC beans include black beans and red haricot, whereas red kidney and pinto fall into the HTC category (and we can even toss in chickpeas). ETC beans need a shorter time to cook because they take up water much faster than HTC varieties. The Cooking Tips section in this chapter presents several methods for cooking beans.

Peas

There are three main types of peas: English peas, snow peas, and sugar snap peas. Only the pea seeds of English peas are edible—their pods are too tough to eat. Snow peas are flat, with underdeveloped peas, and the pods are eaten whole. Sugar snap peas can be eaten whole with their tender pods, and their peas are bigger than snow peas. The young tender leaves and sweet flowers of the pea plant are also edible and great in salads and as garnishes. Dried peas are very popular in Indian cooking; they are used to make dals and dishes like the Goan Pea Curry (page 235).

Peanuts

Peanuts, groundnuts, or monkey nuts are not a nut but a legume and are a popular snack in many parts of the world. In India, they are eaten boiled or roasted. Because they are similar in their composition to nuts, peanuts can be used in various dips, spreads, and sauces, such as the Peanut Muhammara used in the Egg Sandwich (page 275) and the Green Beans + Cucumber Noodles (page 237). When roasted, peanuts take on a wondrous smoky flavor that can be used to build smokiness in various dishes.

Green Beans

This is a broad category of fresh bean pods that are harvested from diverse bean plants when the fruits are immature. There are several varieties, including French beans (haricots verts), snake beans, string beans, runner beans, and snap beans, and they come in a variety of colors.

Chickpeas

I have the impression that chickpeas are the most talked-about legumes. From eating them whole to making spreads and even sweets, chickpeas are an important staple all over the world. Dried chickpeas need some time to cook, so treat them as HTC beans when preparing. If you buy chickpeas from an Indian store or read an Indian cookbook, you will notice two varieties—kabuli and desi. Kabuli are the larger, lighter cream-colored variety and originated in Afghanistan (hence the reference to Kabul in the name) and the Mediterranean. The desi (Hindi for "local") variety is smaller, with a higher fiber content, and the seeds are darker (they come with either green or black skins); these are used for chana dal or split chickpeas. Desi chickpeas are also called Bengal gram or kala chana. In India, chickpea flour is made from desi chickpeas, but the flour from both types can be used interchangeably in most recipes without any noticeable differences. Look for fresh green chickpeas at the farmers' market; they are tender, cook quickly, and are great in salads.

Lentils

Lentils are tiny, round, flat legume seeds that resemble lenses. There are main varieties, including green or le Puy lentils, red lentils, and the black, beluga, or caviar lentils. The seeds do not need to be soaked in water before cooking, and due to their tiny size, they cook quickly. Compared to the other lentils, black lentils take longer to cook and will retain their texture and won't fall apart easily. In India, lentils are a dietary staple and are used to make dals and various stews like sambar with vegetables or meat.

220

Jícama

Jícama, also called Mexican yam bean or Mexican turnip, is a large round tuberous root. The skin is thin like potato's but like a finer tissue paper. The root is rich in water and crisp, with a nutty, pear-like taste and a very mild hint of sweetness. The jícama plant also produces edible peas in pods that look similar to lima beans.

Storage

Fresh peas, beans, and chickpeas should be stored in the refrigerator; dried peas, beans, chickpeas, and lentils must be stored away from light and heat in a cool, dark spot. Peanuts are rich in fat, so they are best stored in the freezer to prevent rancid flavors from developing. Jícama should be stored unwrapped in a cool, dark spot; it can keep for up to 4 months. Peel the skin before use.

Cooking Tips

- Lentils do not need soaking in water prior to cooking but they should be cleaned and rinsed.

- While fresh (and frozen) peas cook quickly, treat dried peas like beans and give them a soak prior to boiling (see Goan Pea Curry, page 235).

- Remove and trim the stem tops and any damaged ends from French beans and other green beans, as well as the thick string that runs along the seam of the bean. Blanch green beans in a pot of salted water for a few seconds until they turn bright green, then immediately plunge them into a bowl of ice water to prevent them from cooking further. Green beans can also be stir-fried, baked, roasted, or grilled.

- If green beans are overcooked, they will be tender but lose their intensity of color and turn dull. If you want to add them to pulao, stews, or braises, add them in the final 10 minutes of cooking.

- Beans can be cooked faster by removing their skins, presoaking, or boiling them in alkaline and monovalent solutions (read on for a description).

- I live in an area where the water is hard and contains a considerable amount of calcium and magnesium salts. Calcium and magnesium (called *divalent salts*—each element contains two exchangeable electrons) can enter the structural carbohydrate pectin molecule, making the seed stubbornly hard and take longer to cook. I use softened, deionized, or filtered water to cook beans. If that's not an option, brine the beans, or boil them in water containing an excess amount of regular salt (sodium chloride) and baking soda (method follows). Monovalent ions (containing a single exchangeable ion) like those in sodium enter the pectin molecule, bumping out the calcium and magnesium and making the pectin softer and faster to cook.

- Pressure cooking beans is a lifesaver. The combination of high pressure and heat in the presence of water helps dried beans cook in a fraction of the usual time, anywhere between 15 and 30 minutes. I still like to soak beans prior to pressure cooking because this helps awaken the enzymes and kick off various plant biochemical pathways that help beans become easier to cook and digest.

Brining HTC beans and chickpeas

Make up a solution of **1 Tbsp [15 g] salt** + **1 tsp [5 g] baking soda to 4½ cups [1 L] water per 1 lb [455 g] beans**. Soak the beans for 16 to 24 hours at room temperature. Drain, rinse, and boil until tender.

Boiling HTC beans and chickpeas

Soak the beans in plain water overnight. Drain. The general ratio for 1 lb [455 g] dried beans is **1 tsp [6 g] salt** + **¼ tsp [1.5 g] baking soda to 4½ cups [1 L] water**; boil in this solution until they become soft and creamy. This method is great when you need the beans to be extremely soft and mushy, like in the dal makhani recipe in *The Flavor Equation* cookbook.

- Beans are notoriously associated with flatulence. We lack the enzyme alpha-galactosidase to break down certain galactose-containing sugars—like raffinose, stachyose, and verbascose—in beans. Our gut bacteria ferment these sugars, which results in bloating and flatulence. Soaking beans in water or brining them helps reduce the occurrence of flatulence by removing these sugars through the action of enzymes present in the beans. Some of those sugars are released into the soaking water, so it should be discarded and fresh water used for cooking. Just a heads-up: None of these methods is 100 percent effective; because all beans are created differently and contain varying amounts of these sugars, they respond differently to these methods. In South Indian cooking, spices like ground fenugreek seeds are added to fermented dosa and idli batters made from rice and legumes; the yeast and bacteria, combined with the fenugreek, help with digestion of these sugars. As always, a candle can be lit—or try Beano and Gas-X (this section is not sponsored by them).

- Kombu or kelp contains alpha-galactosidase, an enzyme that digests raffinose and is used in preparing beans to reduce flatulence. The enzyme works best at an acidic pH and at 140°F [60°C]; it loses its efficacy at temperatures above 176°F [80°C], so adding kombu to boiling water will not be very useful. I soak two 6 in [15 cm] sheets of kombu in 4½ cups [1 L] water at 140°F [60°C], add 1 tsp [6 g] fine sea salt and 1 tsp [5 ml] fresh lemon juice, and maintain the temperature overnight using a sous vide device. If you don't own a sous vide device, an Instant Pot will work: Set it at the warm setting, which ranges from 145°F to 172°F [63°C to 78°C] as per the manual. You can also replace the kombu with a Beano pill, which also contains the enzyme alpha-galactosidase.

- To save time during the week, I cook beans and lentils, drain, and freeze in ziptop bags for up to a month.

- Jícama, like cucumber, releases a lot of water when tossed with ingredients like salt, sugar, and dressings or vinaigrettes. Add it to a salad just before it's ready to be served or eaten.

- Lectins are another hot topic that comes up with beans and cereals like wheat. Lectins are a type of protein found in beans that can bind carbohydrates. Lectins in their native state can bind proteins and are a very useful scientific tool when trying to study blood types; however, this ability to bind carbs can cause health issues by affecting nutrient absorption and a host of other gut-related issues. Foods high in lectins are rarely consumed raw; we don't eat beans raw, and properly hydrating beans by soaking and cooking them significantly destroys the lectins. Removing the seed hull and sprouting beans is another way to reduce the amount of lectin. However, slow-cooking beans by simmering at a very low temperature isn't the most effective way to get rid of all the lectins; the better option is to presoak the beans and cook them at a higher temperature. In short, cook your beans well, and don't eat them raw or undercooked.

- Some peas, like raw sugar snap peas, might feel a bit tough; blanching (30 seconds) or steaming (1 minute) followed by stir-frying will help make them tender.

- A word on canned and precooked (steam-packaged) lentils and beans: If it makes cooking easier, use them. If you want to soak and cook your beans from scratch, do it. The only person who needs to have a say in this is you.

Aquafaba

While I've stayed away from desserts and cocktails in this book, I cannot write this section and leave out aquafaba. Aquafaba—the leftover bean cooking liquid—is a fantastic egg white substitute that foams when whipped (this also reduces waste and conserves water). I find that it is better to use the liquid from canned chickpeas and beans (the consistency and quality are superior to the homemade stuff); the aquafaba from chickpeas and white beans such as navy is neutral in taste and colorless, so it won't add a beany taste to your recipe. This is my general vegan meringue recipe, adapted from my column at *Serious Eats*. Feel free to play with the flavor and colors.

MAKES 4 SERVINGS

Use the **aquafaba drained from a can of chickpeas (about ½ cup [120 ml])** and add **½ cup [100 g] superfine sugar, ¼ tsp cream of tartar or ½ tsp lemon juice**, and **¼ to 1 tsp of any spice and flavoring ingredient**. Whisk using a stand mixer until stiff peaks form, 10 to 12 minutes. Prep the meringue into whatever shapes your heart desires and bake them on baking sheets lined with parchment paper in a preheated oven at 200°F [95°C] for at least 1 hour, until the tops are firm. They won't turn golden brown, as the acid in the cream of tartar or lemon juice will prevent that reaction. Cool completely before eating.

4 Qt

2 Qt

11. Chickpeas, Soybeans, Lentils, Peas, Beans + Jícama

Crispy Spiced Chickpeas

MAKES 2 CUPS [240 G]

Preheat the oven to 300°F [150°C]. Line a rimmed baking sheet with parchment paper or foil.

Toss (it's easier to just shake the sheet on the counter, so the chickpeas roll in the oil and salt) **two 15 oz [425 g] cans chickpeas, rinsed, drained, and patted dry with clean paper towels**, with **2 Tbsp extra-virgin olive oil** and **½ tsp fine sea salt**. Dry the chickpeas in the preheated oven for 1½ hours until crisp, rotating halfway through drying. Some of the chickpeas might split, and that's fine. Remove and let cool to room temperature completely.

Increase the oven temperature to 300°F [150°C]. Return the chickpeas to the oven and dry again for another 30 minutes or until crisp.

Prepare the seasoning mix. In a medium mixing bowl, combine **1 tsp ground cumin, 1 tsp smoked sweet paprika, 1 tsp ground sumac or amchur**, and **¼ to ½ tsp ground cayenne**. Add the hot chickpeas and toss to coat well (you can also mix them together on the baking sheet). Taste and add **fine sea salt** as needed. Let cool to room temperature before storing in an airtight container for up to 1 month.

Whether you decide to eat these as a snack or use them to replace croutons (see Sweet Potato Kale Caesar Salad, page 136), you're making good decisions.

THE COOK'S NOTES

- The chickpeas are first oven-dried and then allowed to cool to room temperature; this allows the chickpeas to absorb moisture from the air. The second oven-drying wicks out that moisture; this helps produce a longer-lasting crispy texture.

- This is a good base recipe that you can tweak with different seasonings such as garam masala, za'atar, and even dried harissa seasonings (not the paste).

- Save and hold on to those little packets of silica gel beads that come with all sorts of dried foods like seaweed, dried shiitakes, snacks, and supplements. Pop a few into the container with these chickpeas; they will help control the moisture and keep the chickpeas crisp.

Jícama + Pea Salad with Yuzu Vinaigrette

MAKES 4 SERVINGS

Fill a medium bowl with ice water. In a small pot over medium-high heat, bring to a rolling boil **2 cups [480 ml] water** and **½ tsp fine sea salt**. Add **1 cup [120 g] fresh or frozen peas**. Cook until the peas turn bright green and tender, about 2 minutes. Strain the peas and transfer them to the ice water to stop the cooking process.

Strain the peas again and add to a large mixing bowl along with **1 lb [455 g] jícama bulb, peeled and cut into ¼ in [6 mm] thick matchsticks, about 1½ in [4 cm] long**; **1 shallot, cut in half and thinly sliced**; **1 large bell pepper, cored and thinly sliced lengthwise**; **1 fresh green chilli, such as a jalapeño or serrano, minced**; and **2 packed Tbsp chopped fresh mint leaves**.

Yuzu Vinaigrette

In a separate small bowl, whisk together **2 Tbsp rice wine vinegar**; **2 Tbsp sesame oil**; **1 Tbsp yuzu ponzu**; **1 Tbsp toasted sesame seeds**; **1 tsp dark brown sugar**; **1 tsp red pepper flakes such as Aleppo, Maras, or Urfa**; and **½ tsp ground black pepper**.

Pour the dressing over the vegetables in the bowl. Toss well to coat. Taste and season with **fine sea salt** as needed. Serve immediately. This salad does not store well.

This is a summery salad teeming with the citrusy fragrance of yuzu, vinegar, and mint. I serve this with seafood and with grilled meats and vegetables. Sometimes, I replace the mint with tarragon. This salad is a wonderful substitute to the poppy seed coleslaw served with the Chicken Katsu (page 182).

THE COOK'S NOTES

- Just like in the Cucumber + Roasted Peanut Salad (page 189), the jícama here is rich in water and will release liquid as soon as it's dressed. Eat this salad within 30 minutes of preparation.

- For a stronger sesame flavor, use toasted sesame oil.

- Ponzu is a citrus-based soy sauce that comes in various flavors. Yuzu ponzu contains the juice from the yuzu citrus for both citrus and floral notes. Try other varieties of ponzu to play with the flavors in the vinaigrette. Use ½ Tbsp low-sodium soy sauce + ½ Tbsp fresh lemon or orange juice + ½ tsp lemon or orange zest as a substitute.

226

Snow Peas with Dukka + Feta

MAKES 4 SERVINGS

To make the dukka, heat a large stainless-steel skillet over medium-low heat. Add **1 cup [140 g] raw nuts such as almonds, hazelnuts, cashews, pine nuts, walnuts, or a mix of nuts**. Toast until the nuts are fragrant and light brown, 3 to 4 minutes. Remove from the heat and transfer to a small plate or bowl and cool completely to room temperature.

Wipe the pan clean with a lint-free kitchen towel and add **1 cup [140 g] white sesame seeds**, **2 Tbsp whole coriander seeds**, **2 Tbsp whole cumin seeds**, and **2 tsp whole peppercorns**. Toast over medium-low heat until the seeds are light brown and fragrant, 3 to 4 minutes. Remove from the heat, add to the bowl of nuts, and let cool completely to room temperature.

Transfer the nuts and seeds to a food processor, coffee mill, or spice grinder and add **2 tsp red pepper flakes such as Aleppo, Maras, or Urfa**. Pulse to a coarse, gritty consistency. Transfer to an airtight container and store until ready to use. This stays good for up to 1 month at room temperature and 3 months in the freezer.

Wipe the pan clean with a lint-free kitchen towel. Set over medium-high heat and warm **2 Tbsp extra-virgin olive oil**. When the oil is hot, add **1 lb [455 g] snow peas** and **¼ tsp fine sea salt**. Sauté until they are tender, bright green, and slightly charred, 5 to 8 minutes. Remove from the heat and toss with **1 tsp fresh lemon or lime juice** and 2 to 3 extremely generous Tbsp dukka.

Transfer to a serving plate and garnish with **¼ cup [30 g] crumbled feta**. Taste and season with **fine sea salt** if needed. Serve warm. Leftovers can be stored in an airtight container in the refrigerator for up to 3 days.

Dukka (or duqqa) is an Egyptian and Middle Eastern spice blend made from toasted nuts and spices. My favorite version comes from the Palestine kitchen of cookbook author Reem Kassis's mother; she graciously shares it with me every year. For those moments when I've used up what Reem graciously sends me, I turn to this recipe. Sautéed snow peas are generously tossed with dukka and dotted with feta.

THE COOK'S NOTES

- You can use any type of nut and even a combination to make the dukka; just be sure to use raw nuts, either with skin on or skinless.

229

White Beans + Mushrooms in Broth

MAKES 4 SERVINGS

In a medium saucepan, warm **2 Tbsp extra-virgin olive oil** over medium heat. When the oil is hot, add **2 shallots, thinly sliced**. Sauté until they start to turn light golden brown; the time will vary. Add **4 garlic cloves, thinly sliced**; **1 or 2 sprigs of fresh thyme**; **1 tsp red pepper flakes such as Aleppo, Maras, or Urfa**; and **½ tsp ground black pepper**. Sauté until fragrant, 30 to 45 seconds.

Add **8 oz [230 g] button or cremini mushrooms, trimmed and thinly sliced**. Sauté until they start to turn light brown and plump, 2 to 4 minutes. Add **1 cup [240 ml] water, ½ cup [120 ml] dry white wine such as sauvignon blanc or pinot gris**, and **1 tsp low-sodium soy sauce**. Bring to a boil over high heat and stir, scraping the bottom of the pan.

Add **one 14 oz [400 g] can white beans, such as cannellini or Great Northern, rinsed and drained**. Return to a boil. Lower the heat to a simmer and cook until the beans are warmed and the mushrooms are cooked all the way through, about 5 minutes. Remove from the heat, discard the thyme sprigs, and portion into shallow bowls. Garnish each bowl with **2 tsp grated Parmesan** (on a Microplane) and **1 tsp chopped chives**. Serve warm. Leftovers can be stored in an airtight container in the refrigerator for up to 3 days.

It's amazing how savory this bowl of beans and mushrooms becomes. Brothy beans deserve—no, they *demand*—warm slices of sourdough bread brushed with butter or olive oil. Make sure you have plenty of bread ready. This is also good served in a bread bowl.

THE COOK'S NOTES

- Clean the mushrooms just before cooking; if you handle them too much before storing, they'll spoil. Brush any dirt off whole mushrooms and rinse them under water to dislodge any dirt. Pat dry with a clean paper towel.

- When prepping mushrooms, trim the end of the stem if it is very woody or dry, but if it is tender, you can leave it.

230

Black Beans, Corn + Gochujang Saag

MAKES 4 SERVINGS

In a medium saucepan, warm **2 Tbsp sesame oil or extra-virgin olive oil** over medium heat. When the oil is hot, add **1 shallot, halved and thinly sliced**, and **2 garlic cloves, thinly sliced**. Sauté until the shallots become translucent, 3 to 4 minutes. Add **1 large red bell pepper, cored and diced**. Sauté until the peppers just start to soften, 4 to 5 minutes. Using kitchen tongs, fold in **10 oz [282 g] baby spinach** a handful at a time as it cooks down. Sauté until the leaves are completely wilted, and continue to cook until the leaves release most of their liquid.

Add **one 14 oz [400 g] can black beans, drained and rinsed**, and **1 cup [150 g] fresh or frozen sweet corn** (no need to thaw). Stir in **1 cup [240 ml] plain, unsweetened full-fat coconut milk**, **¼ cup [80 g] gochujang**, **2 tsp low-sodium soy sauce**, and **½ tsp ground black pepper**. Bring to a boil over high heat. Immediately lower the heat and let simmer for 1 to 2 minutes for the flavors to meld. Remove from the heat, taste, and season with **fine sea salt**.

Transfer to a serving bowl and garnish with **2 Tbsp chopped chives**. Serve hot or warm with rice. Leftovers can be stored in an airtight container in the refrigerator for up to 3 days.

Now, it might seem odd that I've opted to use the Hindi word for spinach in a stew made with Korean gochujang, but let me explain. During my time in grad school in Washington, DC, I once had to cook and bring something to a party, the theme being "Cook Something Based on the Last Dish You Ate." You guessed it: I'd eaten Korean food a few days before, so I used it to concoct this coconut milk stew. It's pure unadulterated comfort food, best eaten with rice.

THE COOK'S NOTES

- The intensity of the red color of the final dish might vary depending on the brand of gochujang you use.

- Thinly sliced scallions are a good alternative to chives.

- Spinach shrinks down dramatically when cooked and releases a lot of liquid. Add the leaves in batches; that will make it easier to manage. The water released contributes to the stewy liquid.

Goan Pea Curry

MAKES 4 TO 6 SERVINGS

In a medium saucepan, bring to a boil over high heat **3 cups [720 ml] water**; **1½ cups [300 g] dried yellow, white, or green peas, soaked overnight, rinsed, and drained**; **1 tsp fine sea salt**; and **⅛ tsp baking soda**. Lower the heat to a simmer, cover, and cook until the peas are completely cooked and tender, 45 to 75 minutes. Remove from the heat and set aside the peas with the cooking liquid. No need to remove the skins that will have risen to the top, but if you must, use a slotted spoon to skim them out.

In a medium skillet, warm **2 Tbsp neutral oil with a high smoke point such as grapeseed** over medium heat. When the oil is hot, add **2 medium yellow or white onions, halved and thinly sliced**. Sauté until they start to brown, 4 to 6 minutes. Add **2 Tbsp peeled and grated fresh ginger** and **6 garlic cloves, grated**. Sauté until warmed and fragrant, 30 to 45 seconds. Add **½ cup [40 g] unsweetened fresh, frozen, or desiccated shredded coconut**. Sauté until the coconut begins to darken slightly, 1 to 2 minutes. Remove from the heat and transfer to a blender or food processor.

To the blender, add **1½ tsp garam masala, homemade (page 341) or store-bought**; **2 Tbsp chopped cilantro**; **1 tsp ground Kashmiri chilli powder (or ¼ tsp smoked sweet paprika + ¼ tsp ground cayenne)**; and **1 cup [240 ml] water**. Pulse until smooth.

Clean the saucepan and warm **2 Tbsp neutral oil with a high smoke point such as grapeseed** over medium-high heat. Add **1 large yellow or white onion, minced**, and sauté until light brown, 4 to 5 minutes. Lower the heat to medium, then stir in the ground coconut mixture and cook until it starts to bubble, 1 to 2 minutes. Stir in **2 Tbsp tomato paste**, **1 tsp sugar**, and **¼ cup [60 ml] water**. Once the mixture begins to bubble again, lower the heat to a simmer. Continue to cook, stirring, until the mixture starts to pull away from the sides of the saucepan and the fat begins to separate, 5 to 8 minutes.

Fold in the peas with the cooking liquid. Taste and season with **fine sea salt**. Bring to a boil over high heat to warm the peas through and then remove from the heat. Garnish with **2 Tbsp chopped cilantro**.

Serve hot or warm with **plain rice or flatbread**. Leftovers can be stored in an airtight container for up to 3 days and frozen for up to 1 month in a freezer-safe airtight container.

continued

This dish comes from the western coastal state of Goa in India and is made from dried peas (fresh peas are too tender to use here). Frankly, I don't like the way the English name of this dish translates—dry pea curry sounds miserable. I prefer referring to the dish by its Goan name, chanya ros (it also goes by vatana curry or chanya ros).

THE COOK'S NOTES

- Dried white, green, or navy peas can be found online or in Indian grocery stores. They will be labeled as vatana or dried peas.

- For a deeper coconut aroma, use coconut oil.

- When boiled, the peas will release their skins, which will accumulate on the surface of the water. You can leave them in if the texture doesn't bother you; otherwise, skim them off with a slotted spoon and discard.

236

Green Beans + Cucumber Noodles

To prepare the sauce, in a small mixing bowl, whisk together **⅓ cup [90 g] smooth, unsweetened peanut butter**; **¼ cup [60 ml] low-sodium soy sauce**; **¼ cup [60 ml] Chinese black vinegar**; **2 Tbsp Shaoxing wine or dry sherry**; **3 garlic cloves, grated**; **2 tsp honey or maple syrup**; and **½ tsp toasted, ground Sichuan peppercorns** (see the Cook's Notes). Don't worry about the peanut butter floating on the liquid; once heated, it will eventually come together.

Bring a large pot of salted water to a boil over high heat and add **12 oz [340 g] dried ramen or chow mein stir-fry noodles**. Cook according to the package instructions. Drain the cooked noodles in a colander, rinse under cold running water, and shake well to get rid of excess water. Leave in the colander. Drizzle with **2 Tbsp toasted sesame oil** and toss to coat and prevent the noodles from sticking together.

In a wok or large, deep stainless-steel skillet, warm **2 Tbsp neutral oil with a high smoke point such as grapeseed** over high heat. Once the oil begins to shimmer, add **8 oz [230 g] green beans or haricots verts, trimmed and cut into 2 in [5 cm] pieces**, and **fine sea salt**. Stir-fry until the beans are bright green and lightly browned, 2 to 3 minutes. Add **4 scallions, both white and green parts, cut into 2 in [5 cm] pieces, any thick bulbs sliced in half lengthwise**. Stir-fry until the scallions are bright green, 1 to 2 minutes. Remove from the heat.

Transfer the noodles to a bowl, add the sauce, and fold to coat well. Taste and season with **fine sea salt**. Top with the greens beans and scallions; **1 medium English cucumber, seeds removed and sliced into ¼ in [6 mm] matchsticks**; **½ cup [70 g] roasted whole salted peanuts, crushed**; and **½ cup [10 g] tightly packed cilantro**. Drizzle with **1 to 2 tsp chili crisp** and serve immediately.

Leftovers don't store well, as the noodles dry out and the fresh cucumber loses its appealing texture after a few hours.

continued

There is a dish that I religiously order at least once a month and that's the dan dan noodles at the Taiwanese restaurant Pine and Crane in the Silverlake neighborhood of Los Angeles. I love it so much that I've also gone there two days in a row just for the noodles. For those moments when it's impossible for me to get there, this satisfies the craving.

THE COOK'S NOTES

- It's best to toast a large batch of Sichuan peppercorns and keep them in your pantry. Toast them in a dry cast-iron or stainless-steel skillet over medium-high heat until the peppercorns start to turn lightly brown and fragrant, 30 to 45 seconds. Transfer to a plate and let cool before grinding them down to a fine powder using a spice grinder or coffee mill. Store the powder in an airtight container for up to 3 months.

- Whether you use commercial or natural peanut butter, remember to stir the peanut butter very well before you use it.

- Once the noodles and sauce are mixed, eat the dish quickly. The noodles suck up the sauce quickly and can turn dry after a few hours.

238

Orecchiette with Spiced Peas

MAKES 4 SERVINGS

Bring a large pot of salted water to boil over high heat and add **1 lb [455 g] dried orecchiette or pasta like farfalle**. Cook until al dente, per the package instructions. Toward the last 2 minutes of cooking, add **10 oz [285 g] fresh or frozen peas (no need to thaw)**. Cook until tender and bright green, 2 minutes. Remove the pasta and peas with a slotted spoon or spider and transfer to a medium mixing bowl. Reserve 1 cup [240 ml] of the pasta water, then drain the pasta. To prevent sticking, toss the pasta and peas in the colander with **1 Tbsp extra-virgin olive oil**.

Wipe the saucepan clean and return to the stove. Over medium heat, warm **3 Tbsp extra-virgin olive oil**. When the oil is hot, add **1½ tsp whole cumin seeds, lightly crushed**, and **½ tsp fennel seeds, lightly crushed**. Sauté until fragrant, 30 to 45 seconds. Add **4 garlic cloves, grated**, and **1 tsp red pepper flakes such as Aleppo, Maras, or Urfa**. Sauté until fragrant and the oil starts to turn red. Stir in **2 Tbsp white or yellow miso paste** and ½ cup [120 ml] of the reserved pasta cooking water. Whisk until smooth. Add the cooked peas and smash with a potato masher or spoon just to break open about half of the peas—don't mash them all to make a paste.

Fold in the cooked pasta and **¼ cup [15 g] grated Parmesan**. Add more pasta cooking water, 1 Tbsp at a time, if needed. The final consistency should be slightly brothy. Taste and season with **fine sea salt** if needed. Garnish with **2 Tbsp chopped chives** and **2 Tbsp grated Parmesan**. Leftovers can be stored in an airtight container in the refrigerator for up to 3 days.

This makes a big sprawling pot of pasta and peas—though peas are a spring vegetable, frozen peas provide the option to make it year-round. I love to garnish the pasta with tender pea shoots when in season.

THE COOK'S NOTES

- Cumin helps accentuate the heat from the red pepper flakes.

- Miso and Parmesan give this a deep edge of savoriness.

Lentil Lasagna

In a large saucepan over high heat, bring to a boil **4 cups [945 ml] water, 1 cup [200 g] black lentils,** and **½ tsp fine sea salt.** Turn down the heat to low and cook until the lentils are completely tender but not falling apart, 25 to 30 minutes. Drain.

Warm **2 Tbsp extra-virgin olive oil** over medium heat in a large pan. When the oil is hot, add **1 large white or yellow onion, cut into ¼ in [6 mm] dice.** Sauté until the onions become translucent, 4 to 5 minutes. Add **2 celery stalks, diced into ¼ in [6 mm] cubes,** and **1 large carrot, trimmed, peeled, and diced into ¼ in [6 mm] cubes.** Sauté until the carrots just start to soften, 3 to 4 minutes. Stir in **one 28 oz [794 g] can crushed tomatoes; 2 large red bell peppers, cored and finely chopped; 2 tsp red pepper flakes such as Aleppo, Maras, or Urfa; 1 tsp garam masala, homemade (page 341) or store-bought; ½ tsp ground turmeric; ½ tsp ground black pepper;** and **1 cup [240 ml] water.** Bring to a boil over high heat. Remove from the heat and use an immersion blender to pulse for a few seconds to get a chunky sauce. (Alternatively, transfer the mixture to a blender or food processor, pulse, and return the sauce to the pot.) Taste and season with **fine sea salt.**

Fold in the drained cooked lentils. Cook over medium heat until most of the liquid evaporates and the sauce is reduced to about 6 cups [1.4 L]. If the mixture feels too runny, thicken by stirring in **2 Tbsp all-purpose flour.**

Bring a large pot of salted water to a boil and add **1 lb [455 g] lasagna sheets** (see the Cook's Notes). Cook until al dente, per the package instructions. Remove and toss with **a little extra-virgin olive oil** to prevent them from sticking.

In a medium bowl, combine **1½ cups [120 g] shredded mozzarella** and **1½ cups [90 g] grated Parmesan.**

Meanwhile, preheat the oven to 350°F [180°C]. Generously grease a 9 by 13 by 2 in [23 by 33 by 5 cm] deep rectangular baking pan with **a little unsalted butter or extra-virgin olive oil.**

continued

Line the base of the prepared baking pan with enough sheets of pasta to cover in a single layer. Cover the pasta with 1½ cups [360 g] of the tomato-lentil mixture using a spoon or an offset spatula. Sprinkle about ¾ cup [160 g] of the cheese mixture and place another layer of pasta on top. Repeat and layer with the tomato-lentil mixture and cheese until all the pasta is used. The final layer should be covered with the tomato-lentil mixture and cheese. Cover the dish tightly with a layer or two of aluminum foil. (If the lasagna goes to the top edge of your pan, place the baking pan on top of a sheet pan to catch any drips.) Bake for 35 to 45 minutes, until the cheese is melted and the sauce is bubbling. If you are using no-boil pasta, it should be cooked through. Remove the pan from the oven and remove the foil. Return the pan to the oven and cook on the upper rack until the cheese begins to bubble and turns light brown, 4 to 5 minutes. Remove from the oven and let cool for at least 10 minutes before serving. Serve warm. Leftovers can be stored in the refrigerator in an airtight container for up to 3 days or frozen in a freezer-safe airtight container for up to a month.

A classic Italian dish and a staple at many potlucks, lasagna is beloved for a reason. In my previous book, *The Flavor Equation*, I included a recipe for dal makhani, a versatile dish that is also great on nachos and in lasagna. This time around, I'm revisiting the idea of lentils in lasagna using a different dal recipe. Prepare to feel satisfied and comforted.

THE COOK'S NOTES

- Save yourself an extra step with no-boil lasagna sheets, which are thinner and partially cooked.

- Black lentils hold their shape here, which is critical to the interplay of textures in this lasagna.

- A common issue with lasagna making is how wet it ends up. To avoid this, thicken the sauce with a little flour and use low-moisture mozzarella.

Indian Lamb + Lentil Stew

MAKES 4 TO 6 SERVINGS

In a medium bowl of cold water, soak **1 cup [200 g] split pigeon peas, picked for debris and rinsed**, for 30 minutes.

While the pigeon peas soak, in a large Dutch oven or saucepan over medium-high heat, warm **¼ cup [60 ml] ghee or extra-virgin olive oil**. When the oil is hot, add **4 large white or yellow onions, chopped**. Sauté until they start to caramelize and turn light brown. If the onions seem dry at any time, add 1 Tbsp water to prevent them from burning, and scrape up any browned bits from the bottom of the pot. This could take as long as 45 minutes. Remove and transfer the onions to a blender or food processor. Add **one 1 in [2.5 cm] piece fresh ginger, peeled and sliced into thin coins**; **4 garlic cloves**; and **¼ cup [60 ml] water**. Purée until smooth.

Clean the Dutch oven, return it to the stove, and heat **2 Tbsp ghee or extra-virgin olive oil** over medium-high heat. Add, in batches if necessary to keep from crowding the pan, **2 lb [910 g] boneless leg of lamb, trimmed of any excess fat, and cut into ½ in [13 mm] cubes**. Sauté until the meat turns brown, 4 to 5 minutes. Remove the meat and transfer to a medium bowl. Leave any liquids in the Dutch oven.

Return the blended onion mixture to the Dutch oven and turn down the heat to low. Stir in **1½ tsp garam masala, homemade (page 341) or store-bought**; **½ tsp ground cayenne**; and **½ tsp ground turmeric**. Cook until almost all the water evaporates and the fat starts to separate from the onions, stirring constantly and scraping the bottom of the pot to avoid burning. The mixture will caramelize and thicken to form a loose paste. Here's another place where it would be unwise for me to predict a time because it will vary considerably. Since the mixture tends to spurt like a hot volcano, use a splash screen and a long-handled spoon or spatula.

Fold in the browned lamb with any accumulated juices; soaked lentils; **2 medium carrots, trimmed, peeled, and diced into ¼ in [6 mm] cubes**; **1 medium sweet potato, peeled and diced into ¼ in [6 mm] cubes**; **2 cups [480 ml] water**; **1 tsp fine sea salt**; and **¼ tsp ground cayenne**. Cover with a lid and cook over low heat until the vegetables, lentils, and lamb are tender and falling apart, 60 to 75 minutes. Stir occasionally to prevent burning.

continued

Remove from the heat and stir in **2 Tbsp fresh lemon or lime juice**. Garnish with **2 Tbsp chopped cilantro**.

Serve hot with **plain rice or flatbread**. Leftovers can be stored in an airtight container for up to 3 days or in the freezer in a freezer-safe airtight container for up to 1 month.

A soft and creamy texture permeates this stew of tender lamb and lentils. This dish comes from the eastern state of Bengal in India; it is also known as dal gosht, and the traditional choice of meat is mutton. Either split pigeon peas or chana dal can be used here. I like to add carrots and sweet potatoes to the stew to make it heartier, but you can skip them.

THE COOK'S NOTES

- Onions make this sauce—you need to use a large amount. To get the right color, it is critical to caramelize the onions properly until toffee brown and to also slowly cook the spice paste until the fat starts to separate from the mixture. This latter step is an important technique in many Indian recipes, and it helps develop a robust flavor.

- Split pigeon peas originated in India; they also go by toor, tur, tuvar dal, or arhar dal. They can be found in Indian grocery stores as well as online.

247

Okra

The Mallow or Cotton Family

MALVACEAE

Origins

OKRA ORIGINATED IN AFRICA.

Okra

Okra, lady's fingers, or bhindi is a spectacular vegetable that occupies a prominent position in the cuisines of Africa, the American South, and India. Okra is wonderful when fried crisp, roasted, or tenderized in a stew like gumbo. When okra isn't in season, I pursue my love of okra through the bags of frozen chopped okra in the grocery freezers.

Storage

Pat fresh okra dry and store it in loosely sealed plastic bags in the refrigerator.

Cooking Tips

- Okra produces a sticky mucilaginous substance when cut. When added to curries, gravies, and stews, this substance is a boon because it helps thicken the liquid.

- Prior to cutting okra, wash the okra well and pat it completely dry. Water draws out the mucilaginous substance, and pieces of okra might end up sticking to your knife and cutting board. Some cooks like to salt their okra to draw out the mucilage, but I've found this method to be a waste of time and inefficient.

- Larger okra is wonderful for roasting, stuffing, or in Okra Tacos (page 255) while the smaller ones are better for curries (Okra + Shrimp Coconut Curry, page 261) and most other purposes.

- Okra benefits from sour flavors. When stir-frying or sautéing okra for a dry dish, skip liquid cooking acids like lemon juice and use dry ground amchur (sun-dried unripe mango powder, found in Indian stores). Sumac is another great option; however, because sumac can quickly burn and turn bitter, I prefer to add it after the okra is cooked.

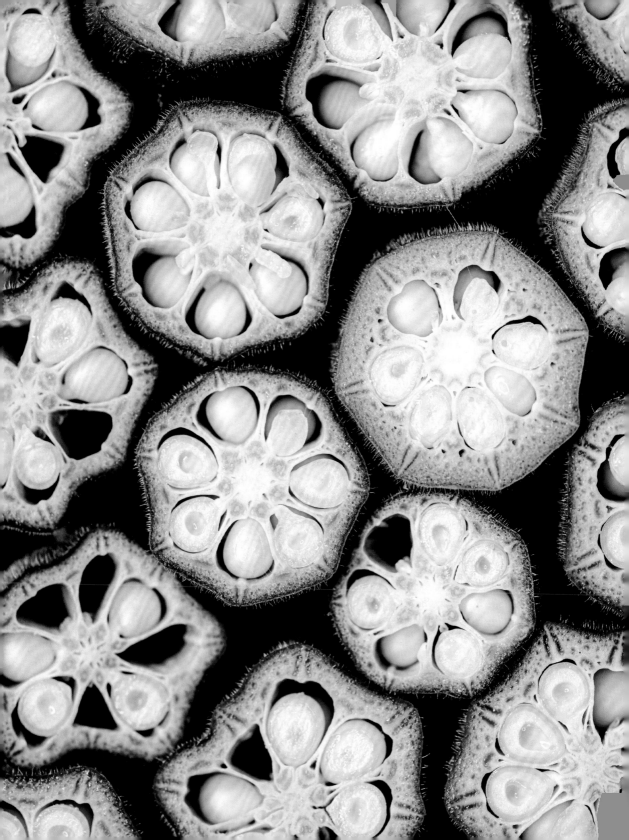

Okra Preserved Lemon Tempura with Tomato Chutney

MAKES 4 SERVINGS

Tomato Chutney

Preheat the oven to 425°F [220°C]. On a baking sheet or roasting pan, toss together **1 lb [455 g] cherry tomatoes**; **2 shallots, halved lengthwise**; and **2 Tbsp extra-virgin olive oil**. Roast until the tomatoes burst and the shallots and tomatoes turn a light golden brown, about 25 minutes. Remove from the oven and transfer the vegetables with the pan juices to a blender.

To the blender, add **4 garlic cloves**, **1 fresh chilli such as jalapeño or serrano**, and **2 Tbsp apple cider vinegar**. Blend until smooth. Taste and season with **fine sea salt** as needed. Transfer to a serving bowl.

In a small saucepan over medium heat, warm **2 Tbsp neutral oil with a high smoke point such as grapeseed**. When the oil is hot, add **1 tsp black or white sesame seeds and 10 to 12 fresh curry leaves (optional)**. Fry until the seeds sizzle and turn fragrant and the leaves turn crisp, 30 to 45 seconds. Pour the hot flavored oil and spices over the tomato chutney. Serve warm or at room temperature. The chutney can be made a day in advance and tastes good for up to 3 days if stored in an airtight container in the refrigerator.

To prepare the tempura, in a medium cast-iron or stainless-steel pan, heat **4 cups [945 ml] neutral oil with a high smoke point such as grapeseed** to 350°F [180°C].

While the oil warms, take two bowls, one that can sit comfortably inside the other, and fill the larger bowl with ice-cold water. In the smaller bowl, whisk together **1 cup [120 g] store-bought tempura batter mix** and **¾ cup [180 ml] chilled water**. The batter should be smooth and free from lumps. Fold in **8 oz [230 g] okra, thinly cut lengthwise on the diagonal**, and **1 whole preserved lemon, rind rinsed, drained, and cut into matchsticks, flesh discarded**. Toss to coat in the batter.

continued

Line a large plate or baking sheet with absorbent paper towels or a wire rack to collect the excess oil. Fry the coated vegetables in batches, adding a few to the hot oil at a time and stirring to separate before adding more, until crispy and golden, 2 to 4 minutes. (Let the oil return to 350°F [180°C] between batches before adding the next.) Transfer with a slotted spoon to the prepared plate.

Season with **fine sea salt**. Serve hot or warm with the tomato chutney. Leftovers can be refrigerated for up to 3 days in an airtight container. Reheat the okra in an oven at 300°F [150°C].

There are lots of unexpected ingredients and surprises in every bite of this dish. Okra and slices of preserved lemon peels coated in tempura batter are fried until crispy and then served with my rendition of a South Indian tomato chutney. The chutney is sweet and tangy with a hint of heat. Be warned, you'll want to eat it with everything, all the time. The preserved lemon peel leaves a delicious tangy flavor that brightens the okra.

THE COOK'S NOTES

- Do not use fresh lemon peel as a substitute for preserved lemons in this recipe, or they will make you bitter (it will literally leave a bitter taste in your mouth) and the tempura unhappy.

- Make sure the washed okra are completely dry before you start cutting them. The more water they meet, the stickier they get. Slice the okra as thin as possible and they will cook much faster.

- Avoid using too much batter when frying. The coating should be thin to create a crisp shell.

- A reminder when working with preserved lemons: They're made using salt and, consequently, are very salty. Rinse the peels well before use and discard the pulp.

- When making tempura batter, use ice-cold water. This reduces the formation of gluten and produces a crunchier texture.

Okra Tacos with Butter "Chicken" Sauce

MAKES 4 SERVINGS

First, prepare the butter "chicken" sauce. In a large saucepan or Dutch oven over medium-high heat, melt **3 Tbsp unsalted butter**. Add **1 large white or yellow onion, finely minced**. Sauté until translucent, 4 to 5 minutes. Add **4 garlic cloves, grated**, and **2 Tbsp peeled and grated fresh ginger**. Sauté until fragrant, 30 to 45 seconds.

Add **1 tsp garam masala, homemade (page 341) or store-bought**; **½ tsp ground cumin**; **½ tsp ground turmeric**; and **¼ tsp ground cinnamon**. Sauté until fragrant, 30 to 45 seconds. Deglaze the pan as necessary with 1 Tbsp water at a time to avoid scorching. Turn down the heat to low and add **¼ cup [60 g] tomato paste** and **½ to 1 fresh chilli such as jalapeño or serrano, chopped**. Sauté, stirring often, scraping the sides and bottom of the saucepan, again deglazing the pan as necessary to avoid burning, until the tomato paste just starts to darken, 3 to 4 minutes.

Stir in **⅓ cup [80 ml] water** and **1 Tbsp fresh lemon or lime juice**. Increase the heat to medium and cook so the flavors meld, 2 to 3 minutes. Transfer to a blender and add **¼ cup [60 g] plain, unsweetened full-fat Greek yogurt**. Blend on high speed until smooth and velvety. Taste and season with **fine sea salt**. Transfer to a serving bowl.

To prepare the okra, rinse **1 lb [455 g] okra, cut in half lengthwise**. Pat completely dry with paper towels and place in a large mixing bowl. Sprinkle with **1 Tbsp ground coriander**; **1 tsp ground cumin**; **1 tsp garam masala, homemade (page 341) or store-bought**; and **½ tsp Kashmiri chilli powder**. Toss to coat well.

In a large cast-iron or stainless-steel skillet over medium-high heat, warm **2 Tbsp neutral oil with a high smoke point such as grapeseed**. When the oil is hot, add **4 shallots (total weight about 7¾ oz [220 g]), thinly sliced**, and cook until they start to turn golden brown, 4 to 5 minutes. Add the okra and **½ tsp fine sea salt**. Cook, uncovered, until the okra starts to char, 5 to 6 minutes, stirring occasionally and deglazing the pan as necessary with 1 Tbsp water at a time to avoid burning. Lower the heat to medium, cover the skillet with a lid, and cook until the okra starts to turn tender, stirring occasionally, 5 to 6 minutes. Remove from the heat and sprinkle with **1 Tbsp amchur or fresh lemon or lime juice**. Taste and season with **fine sea salt**.

continued

Warm **eight to twelve 6 in [15 cm] diameter corn or flour tortillas** by placing them, one by one, directly on the stove burner until brown and puffy, then flip. Wrap the warm tortillas in a lint-free kitchen towel to keep warm.

Serve the okra with the butter "chicken" sauce and the warm tortillas on the side. Sprinkle the okra with **½ cup [60 g] crumbled feta or cotija** and **¼ cup [5 g] tightly packed cilantro leaves**. Quarter **2 limes** and serve alongside. Leftovers can be refrigerated for up to 3 days in an airtight container.

There is no chicken in this butter "chicken" sauce. I've included it here for two reasons: One, the sauce is delicious, and two, it is versatile. In this Indo-Tex creation, the okra is spiced and cooked to develop smoky flavors. Then it's assembled with a tortilla to form a taco and slathered generously with the butter "chicken" sauce.

THE COOK'S NOTES

- Amchur is made by grinding sun-dried pieces of unripe mango to a fine powder; it is an excellent source of acid. It is sold in Indian grocery stores and online.

Okra, Feta + Barley Salad with Pumpkin Seed Sauce

MAKES 4 SERVINGS

In a medium saucepan, combine **1 cup [200 g] pearl barley, rinsed**; **4 cups [945 ml] water**; **2 large garlic cloves, smashed**; and **1¼ tsp fine sea salt**. Bring to a boil over high heat, then lower the heat to a simmer and cook, covered, until the barley becomes plump and tender, 25 to 30 minutes. Remove from the heat and drain. Discard the garlic and place the barley in a large serving bowl. Add **1¾ oz [50 g] drained oil-packed sun-dried tomatoes, cut into strips**; **½ cup [60 g] crumbled feta**; **½ tsp ground black pepper**; and **1 Tbsp oil from the sun-dried tomatoes**. Taste and season with **fine sea salt**.

In a large bowl, toss **1 lb [455 g] okra, cut in half lengthwise**, with **2 Tbsp neutral oil with a high smoke point such as grapeseed**, **1 Tbsp sesame seeds**, and **1 tsp ground coriander**.

Heat a 12 in [30.5 cm] cast-iron or stainless-steel skillet over high heat. Add the okra and cook, stirring occasionally, until it starts to char and turn tender, 10 to 13 minutes. Remove from the heat and place over the barley.

Pumpkin Seed Sauce

To make the dressing, in a medium dry stainless-steel skillet over medium heat, toast **½ cup [70 g] pumpkin seeds** until light brown, 2 to 3 minutes. Transfer to the bowl of a food processor and add **½ cup [120 ml] boiling water**, **¼ cup [60 ml] neutral oil**, **¼ cup [60 ml] rice wine vinegar**, **2 Tbsp sambal oelek**, and **2 tsp maple syrup or honey**. Blend on high speed until smooth and thin with additional water as needed, 1 Tbsp at a time. Taste and season with **fine sea salt**.

Drizzle 2 to 3 Tbsp of the dressing on top and garnish with **2 Tbsp cilantro leaves**. Serve the okra warm with the extra dressing on the side. Leftovers can be refrigerated for up to 3 days in an airtight container.

Charring is a wonderful way to cook okra, and the bitter-sweet flavors that arise give it a smoky flavor. The okra is topped with tender roly-poly grains of pearl barley and served with a hot, sweet, and tangy pumpkin seed sauce.

THE COOK'S NOTES
- Sambal oelek is a tart, hot chilli paste from Indonesia. It can be found in grocery stores, in Asian markets, and online.
- Boiling water helps hydrate the pumpkin seeds, making them easier to grind.

258

Okra + Shrimp Coconut Curry

MAKES 4 SERVINGS

In a large saucepan or Dutch oven, warm **2 Tbsp extra-virgin olive oil, coconut oil, or ghee** over medium heat. Add **1 large white or yellow onion, halved and thinly sliced**, and sauté until it starts to brown, 5 to 7 minutes. Add **4 garlic cloves, minced**, and **1 Tbsp peeled and grated fresh ginger** and cook until fragrant, 1 minute. Add **10 to 12 fresh curry leaves; 1 tsp garam masala, homemade (page 341) or store-bought; one 2 in [5 cm] piece cinnamon stick; ½ tsp ground turmeric; and 1 tsp Kashmiri chilli powder (or ¾ tsp smoked sweet paprika + ¼ tsp cayenne)**. Cook until fragrant, 30 to 45 seconds. Lower the heat to medium-low.

Stir in **¼ cup [60 g] tomato paste**. Cook until the tomato starts to lightly brown, stirring occasionally, 3 to 4 minutes. Add **¼ cup [60 ml] water**, scraping the bottom of the pot to prevent the spices and tomato paste from scorching.

Fold in **8 oz [230 g] medium-size okra, cut into ½ in [13 mm] slices**. Cover with a lid and cook until the okra is tender, 8 to 10 minutes, stirring often to prevent burning, adding another 1 to 2 Tbsp water as needed. Add **½ pint [170 g] cherry or grape tomatoes, halved; 2 cups [480 ml] unsweetened coconut milk; and 1 lb [455 g] peeled and deveined shrimp, with or without tails left on**. Cook, covered, until the shrimp turn pink, 3 to 5 minutes. Stir in **2 Tbsp fresh lime or lemon juice**. Taste and season with **fine sea salt**.

Remove from the heat and garnish with **2 Tbsp chopped cilantro leaves**. Serve hot or warm. Leftovers can be refrigerated for up to 3 days in an airtight container.

In summer, I crave and eat as much okra as I can. It's a short season, and I need to get as much of this vegetable as I can. My mother also shares my love for okra, and while she doesn't like to cook a lot, this is one of the few dishes that she makes, and I adore it. Slices of okra are stewed together with plump succulent pieces of shrimp in a rich, coconut curry sauce. Serve it with plain boiled rice.

THE COOK'S NOTES

- Since this is a curry and liquid-based, frozen okra will work just fine here (for those times when it's out of season).

261

Bell Pepper Eggplant Potato Tomato + Tomatillo

The Potato or Nightshade Family

SOLANACEAE

Origins

BELL PEPPERS HAIL FROM MEXICO AND CENTRAL AND SOUTH AMERICA. POTATOES AND TOMATOES COME FROM THE ANDES. EGGPLANT ORIGINATED IN INDIA, AND TOMATILLOS IN CENTRAL MEXICO.

Bell Peppers

Bell peppers, capsicum, or sweet peppers are the nonpungent variety of peppers. They come in a huge selection of shapes, sizes, and colors: purple, red, green, yellow, orange, brown or chocolate, and even black (an intensely dark purple). Bell peppers are lovely raw and several times richer in vitamin C than oranges. Roasted bell peppers become smoky and are great for barbecuing in kebabs and to make sauces (Garlic Miso Steak with Roasted Bell Pepper Sauce, page 299), dips (Peanut Muhammara, page 275), and salsas.

Eggplant

Eggplants, aubergines, or brinjals are botanically classified as berries. They come in large and baby sizes, from round globe shapes to elongated Japanese varieties. The colors vary from green and white to purple. The skin is smooth and shiny, and should be firm and never soft. When the spongy flesh is cooked, the texture becomes soft and meaty. This latter feature is a main reason why eggplant is used to replace meat in many dishes.

Potatoes

Potatoes are round or oblong and come in a variety of sizes and colors, including purple, red, yellow, and white. Fingerlings and baby new potatoes are two of my favorites when I want to cook potatoes quickly. Note: Sweet potatoes do not belong to the Solanaceae family, but instead to the Morning Glory or Convovulaceae family of plants (see page 134).

Tomatoes

Tomatoes are best eaten in summer when they reach their full potential. They are sweet, tart, and salty, with a good deal of umami. They come in various colors just like bell peppers (purple, red, green, yellow, orange, brown or chocolate, black, striped zebra varieties—honestly, the options are endless) and in various shapes and textures. Fresh tomatoes are great in salads as well as cooked preparations.

Tomatillos

Tomatillos or Mexican husk tomatoes are harvested early, so they are tangy and tart in taste and will remind you of a green tomato. Ripened tomatillos are sweet in taste because their sugars are allowed to develop. Tomatillos are roasted and used to make salsas. Their high pectin content helps sauces and soups thicken with ease. Wash well before use because they are slightly sticky. I use unripened tomatillos for their tart flavor in the salsa for Crispy Cauliflower with Tomatillo Salsa Verde (page 267).

Storage

Eggplants are best eaten the day you harvest or purchase them. This vegetable perishes very quickly and lasts for a very short time in the refrigerator; you're better off keeping it in a cool spot away from sunlight. Store potatoes in a paper bag in a cool, dark spot. If the skin starts to develop sprouts, remove and discard these before cooking. If the sprouts are too big and the potato becomes soft, discard it—or plant it in the soil to reap a crop of potatoes in a couple of months. Store fully ripe tomatoes in the refrigerator in a paper bag; leave unripe tomatoes out on the kitchen counter until they ripen. Tomatillos follow a similar storage guideline, except that some dishes call for unripened ones; for these, go ahead and store in the refrigerator in a paper bags. However, to ripen tomatillos, leave them out on the kitchen counter for a few days.

Cooking Tips

- After roasting bell peppers, let the peppers cool and then remove as much of the charred skin as possible. Leave some of the burnt skin behind to get a smoky flavor. The skin releases from the flesh of the pepper as it cools.

- Eggplants used to be bitter, but most varieties today have the bitterness bred out. If your eggplants are bitter, try salt; it doesn't remove bitterness but does a good job of masking it. Sprinkle salt on the eggplant slices, let rest for 15 to 20 minutes, rinse, and pat dry.

- For cooking purposes, potatoes are divided into waxy or starchy (floury) varieties depending on the type of starch present. Waxy varieties contain less starch and more water, while the starchy types are richer in starch and low in water.

- Starchy potatoes are great for making French fries, crispy potatoes, gnocchi, and mashed potatoes (Bombay Potato Croquettes, page 266). Starchy potatoes like russet (russets grown in Idaho are called Idaho potatoes) are also low in sugar and don't brown as much on frying. They also absorb flavors like a thirsty sponge, so watch your seasonings!

- Waxy potatoes like the red bliss do not do as well in these recipes because as they heat, they release their water; imagine a French fry that's crispy on the outside but hollow on the inside. New potatoes and fingerlings are waxy potatoes, and when cooked, they hold their shape well and their bite is firmer. Use them in recipes where the potatoes need to hold their structure, like in gratins and potato salads.

- The third type of potato is what I call the "in-betweeners" or all-purpose—they lie somewhere in between waxy and starchy potatoes with medium starch and moisture content; Yukon golds fall into this group.

- For potato soups where the potatoes need to stay in chunks and hold their shape, use waxy potatoes; in soups that are blended and smooth, use a starchy potato (see Leek, Potato + Pancetta Soup, page 277).

- To make crispy potatoes with a creamy interior, boil the potatoes first in salted water and next rough with a fork as the British cookbook author Jane Grigson recommends in her eponymous *Vegetable Book*,

or use J. Kenji López-Alt's method from *Serious Eats*, in which the potatoes are boiled in a mixture of salt and baking soda. Salt and baking soda help soften the pectin between the cells of the potato, which helps release some of the starch granules. The potatoes are drained and then shaken gently to coat them in the loosened starch before baking. The starch forms a crispy coat. I've applied this method to the Cassava Bravas (page 212) to get an extra-crispy coat.

- When white potatoes and eggplants are cut, they turn brown due to the enzyme polyphenol oxidase (see page 114). To prevent this, submerge the cut pieces in a bowl of cold water until ready to use. If you need to keep them in water for a long period, add 2 Tbsp lemon juice. Be sure to drain and pat them dry before using.

- When roasting whole potatoes, there's no need to peel them. The skin is like that of a sweet potato and is breathable. The potato won't burst! However, vegetables like tomatoes and eggplants that have waterproof raincoat-like skins should be pricked (skip this for smaller tomatoes) to let the juices release easily as they heat.

- Tomatoes take to roasting well. Their flavors become more concentrated as the water content reduces through evaporation and their sugars caramelize.

- Cooked tomatoes will often need a pinch of salt and/or sugar to balance their taste. Taste them at the end and judge for yourself.

- There is no shame in using tomato paste and canned tomatoes. Because they're made from

tomatoes that are picked when ripe, their flavors are often deliciously intense. Buy a tube of tomato paste, which is easier to store and use. Tomato paste is a great option for dishes that need a concentrated flavor when you don't have the time to spend hours concentrating the tomato down (Nopalito + Chickpea Coconut Curry, page 326). Canned crushed tomatoes are a great option when fresh are unavailable (Cauliflower Bolognese, page 180). Do what works for you!

- When using fresh tomatoes for a sandwich, I prefer them to be firm and slightly unripe or they collapse and fall apart. Overripened mealy textured tomatoes are better off in a cooked application like a blended soup, because their flavor isn't the best and the recipe's spices and condiments can help rectify that.

- Avoid refrigerating fresh tomatoes when storing. At 40°F [4°C], refrigeration stops the enzymes from producing one of their key aroma molecules, (Z)-3-hexenal. If you did do this, leave the tomatoes out on the kitchen counter for a day or two to help those enzymes kick back in and produce that lovely tomato aroma.

- Remove the papery husks from tomatillos before cooking them or else they burn and make a mess. Tomatillos should be roasted, grilled, or broiled to bring out their best flavors and reduce their tartness.

Bombay Potato Croquettes

MAKES 4 SERVINGS (16 CROQUETTES)

Preheat the oven to 425°F [220°C].

Individually wrap **1½ lb [680 g] starchy potatoes such as russets** with foil and place them on a baking sheet. Bake until tender, about 1 hour.

While the potatoes bake, prepare the black garlic crème fraîche dip. In a small bowl, whisk together **¼ cup [60 g] crème fraîche**; **1 Tbsp lemon juice**; **20 black garlic cloves (¾ oz [20 g]), smashed to a paste** (see the Cook's Notes); **1 Tbsp rinsed, drained, and minced preserved lemon peel**; and **1 Tbsp chopped dill, leaves and stems**. Taste and season with **fine sea salt**. Keep refrigerated until ready to eat.

Now back to the potatoes. Unwrap the foil and let the potatoes rest until cool enough to handle. Peel and discard the skin. Grate the potatoes or pass them through a ricer set over a large bowl. Sprinkle with **1 Tbsp cornstarch** and add **1 large egg yolk**. Fold with a fork to combine uniformly. Add **1 cup [60 g] grated Parmesan**; **1 fresh green chilli such as jalapeño or serrano, minced**; **2 Tbsp chopped cilantro**; **1½ tsp fine sea salt**; **1 tsp garam masala, homemade (page 341) or store-bought**; **½ tsp garlic powder**; **½ tsp ground turmeric**; **½ tsp red pepper flakes such as Aleppo, Urfa, or Maras**; and **½ tsp ground coriander** and fold to combine uniformly. You should get about 1.4 lb [620 g] of the potato mixture. Divide into 16 equal parts by weight and shape each into a ball about the size of a golf ball. If the mixture is sticky, grease the palms of your hands with a **neutral oil with a high smoke point such as grapeseed**. Place the balls on a plate.

In a 4 qt [3.8 L] cast-iron Dutch oven or heavy-bottomed saucepan, heat **4 cups [945 ml] neutral oil with a high smoke point such as grapeseed** over medium-high heat until the oil reaches 350°F [180°C].

Line a large plate or baking sheet with absorbent paper towels or a wire rack to collect the excess oil. Fry three or four balls in batches in the hot oil until golden brown and crispy and they reach an internal temperature of 165°F [75°C], 2 to 3 minutes. Allow the oil to return to temperature before adding the next batch and repeat with the remaining balls. Transfer with a slotted spoon to the wire rack to drain the excess oil.

Serve the hot potato croquettes with the lemon crème fraîche dip and **Tomato Chutney (page 253)**. These are best eaten hot but can be stored in an airtight container in the refrigerator for up to 3 days; rewarm in the oven at 300°F [150°C].

266

The potatoes are first baked in the oven until tender and passed through a ricer to get a fine texture. They are then spiced with garam masala and fried until golden crisp. The cool black garlic crème fraîche sauce is wondrous against the hot crispy texture of the golden orblike croquettes. On a side note, I struggled writing the serving size because I know this is something you and I don't really want to share.

THE COOK'S NOTES

- A heads-up: the 20 garlic cloves listed isn't a mistake. Depending on the brand and what kind of garlic you use, you will either have 1 large clove (see pearl garlic, page 33) or about 10 smaller cloves per head.

- Mix the mashed potatoes well, which is essential for the mixture to hold its shape during frying.

- Black garlic is very sticky. The easiest way to prep it for the dip is to place it on a cutting board and smash it with the blunt end of a knife until it transforms into a smooth paste.

13. Bell Pepper, Eggplant, Potato, Tomato + Tomatillo

Potato + White Bean Salad with Zhug

MAKES 4 SERVINGS

Fill a large saucepan with enough water to cover **1 lb [455 g] baby new potatoes** and stir in **1 tsp fine sea salt**. Bring the water to a rolling boil over high heat. Lower the heat and simmer until the potatoes are tender, 15 to 20 minutes. Drain the potatoes and place them in a large bowl.

Zhug

While the potatoes boil, make the zhug. In a food processor, pulse **1 bunch [115 g] cilantro, leaves and tender stems**; **1 bunch [130 g] flat-leaf parsley, leaves and tender stems**; **3 jalapeños or 2 serranos, stemmed**; **4 garlic cloves**; **1 tsp ground cumin**; **½ tsp ground green cardamom**; and **2 Tbsp fresh lemon juice** for a few seconds to get a coarse mixture. Taste and season with **fine sea salt**. Stir in **¼ cup [60 ml] extra-virgin olive oil**. Reserve 1 cup [200 g] of the zhug for the salad and store the rest for another use.

Come back to the potatoes. Coarsely mash the potatoes with a fork. Add to the potatoes **one 14 oz [400 g] can white beans, such as Great Northern or cannellini, drained**, and **1 shallot, finely minced**. Fold the zhug into the potatoes carefully. Taste and season with **fine sea salt**. Drizzle with **2 Tbsp extra-virgin olive oil** and sprinkle with **2 Tbsp toasted or raw pumpkin seeds, salted or unsalted**. Serve immediately. Leftovers can be stored in an airtight container in the refrigerator for up to 3 days.

Zhug is a fresh, fragrant herb condiment from Yemen that also goes by the name sahawiq. It's hot and spicy, and in this potato and bean salad, the zhug breathes a wave of herby heat. The shallot adds a little bit of crunchiness to contrast with the softer textures of the potatoes and beans. It would be ridiculous to give you quantities for just enough zhug for this dish, because you can and will want to use it elsewhere (with eggs, sandwiches, fish, meat, vegetables, and so on). You can store the sauce in an airtight container in the refrigerator for up to 4 days or freeze it for up to 2 weeks.

THE COOK'S NOTES

- For fresh herbs, trim off the bruised browned ends of the stems and remove any leaves that look sad.

- Potatoes absorb water with the ease of a sponge. Serve this salad as soon you add the zhug.

Gazpacho

To make it easier on the blender, add in the following order: **1 lb [455 g] ripe tomatoes, quartered**; **1 medium English cucumber, peeled, seeded, and diced**; **1 medium red bell pepper, cored and diced**; **4 garlic cloves, chopped**; **2 slices stale white bread, torn**; **1 cup [240 ml] water**; **⅓ cup [80 ml] extra-virgin olive oil**; **3 Tbsp fresh orange juice**; **2 Tbsp champagne or sherry vinegar**; and **½ tsp ground black pepper**. Blend on high speed until silky smooth. If necessary, strain through a fine mesh strainer over a large bowl. The final volume should be about 6 cups [1.4 L]; add more water if needed. Taste and season with **fine sea salt**. Transfer to an airtight container. Store until chilled, preferably overnight for the best flavor.

In a bowl, combine **1 medium green, orange, or yellow bell pepper, cored and finely diced**; **1 medium tomato (about 4¼ oz [120 g]), finely diced**; **1 small English cucumber, seeded and finely diced**; and **1 shallot, finely minced**. Divide the gazpacho among bowls and top with a sprinkling of the vegetables for garnish.

Leftover gazpacho can be stored in an airtight container in the refrigerator for up to 4 days, while the garnish will taste fresh for up to a day.

Gazpacho is the soothing relief to the burn of blistering heat on a hot summer day, and every year without fail, my husband, Michael, will place a request for a large jug of gazpacho. It also goes without saying that it tastes best when made with summer's best, freshest, and ripest produce. The orange juice and champagne vinegar round out the flavor of the soup, making the taste much smoother.

THE COOK'S NOTES

- Any type of white bread works here. If you don't have stale bread on hand, dry out the bread in an oven preheated to 200°F [95°C] until dry and lightly golden brown, about 1 hour. An easier option is to use ½ cup [70 g] plain dried bread crumbs (avoid the fancier seasoned ones).

- Use the ripest tomatoes, bell peppers, and cucumbers for the best flavor.

- Use red bell peppers to make the gazpacho but go with any other color for the garnish so it shows against the red backdrop.

Eggplant in Tomato Curry

MAKES 4 SERVINGS

In a large saucepan, warm **¼ cup [60 ml] ghee or extra-virgin olive oil** over medium-high heat. Add **1 large yellow or white onion, finely diced**, and sauté until translucent, 4 to 5 minutes. Add **4 garlic cloves, chopped**; **1 tsp nigella seeds**; **1 tsp ground coriander**; **1 tsp red pepper flakes such as Aleppo, Maras, or Urfa**; and **½ tsp ground turmeric** and sauté until fragrant, 30 to 45 seconds. Fold in **1 medium globe eggplant, cut into ¼ in [6 mm] cubes**, and sauté until coated with the mixture. Fold in **1 lb [455 g] medium or large tomatoes, quartered**, and **¼ tsp fine sea salt**.

Cover with a lid and simmer until the tomatoes are falling apart and the eggplant is completely tender, 20 to 30 minutes. Stir occasionally to prevent burning. Stir in **2 Tbsp fresh lemon or lime juice** and **1 tsp dried mint**. Taste and season with **fine sea salt**. Garnish with **2 Tbsp chopped cilantro**.

Serve warm with flatbread or rice. Store leftovers in an airtight container in the refrigerator for up to 3 to 4 days.

Though it works both ways, I think of this as more of a breakfast dish than a dinner dish. My earliest recollection of this eggplant curry comes from visiting family in the southern state of Karnataka in India. Chunks of eggplant were stewed in a rich tomato gravy and served with buttery, flaky parathas and a side of plain yogurt at breakfast. Don't skip the nigella.

THE COOK'S NOTES

- While I prefer the nutty flavor of ghee in this recipe, you can choose olive oil, sesame oil (don't use toasted sesame oil), coconut oil, or even grapeseed oil.

- To make this hotter, either use a hotter red pepper flake or add ground cayenne.

Peanut Muhammara Egg Sandwich

Peanut Muhammara

Char **4 large red bell peppers** directly over a gas burner on medium-high heat, turning them with a pair of tongs and rotating occasionally, until the skin develops char marks and the flesh turns soft, 12 to 15 minutes. Alternatively, place them on a foil-lined baking sheet on the upper rack of the oven and broil on high, rotating them occasionally, for 12 to 15 minutes. Let the bell peppers rest on a cutting board. Once they are cool enough to handle, remove most of the burnt skin and discard the stalk and seeds. Transfer the bell peppers and the remaining charred skin to a blender or food processor.

To the blender, add **1 cup [140 g] roasted unsalted whole peanuts**; **2 Tbsp extra-virgin olive oil**; **2 Tbsp pomegranate molasses**; **2 Tbsp smoked sweet paprika**; **1 Tbsp red pepper flakes such as Aleppo, Maras, or Urfa**; **1 garlic clove**; **1 tsp fresh lemon juice**; and **¼ tsp ground cayenne (optional)** and blend until smooth. Taste and season with **fine sea salt**. Drizzle with **a little extra-virgin olive oil** and store in an airtight container in the refrigerator for up to 1 week.

Fill a medium bowl with ice water and set aside. Place **4 large eggs** in a medium saucepan and fill with cold water to cover the eggs by 1 in [2.5 cm]. Bring the water to a boil over medium heat, and immediately remove from the heat. Let sit for 5 minutes, remove the eggs carefully with a slotted spoon, and leave them in the ice-water bath to stop them from cooking further. Peel the eggs under running water and cut each into 4 slices lengthwise.

In a large mixing bowl, toss **1 cup [20 g] packed arugula leaves** with **1 Tbsp extra-virgin olive oil, 1 tsp fresh lemon juice, ½ tsp ground black pepper**, and **fine sea salt**.

To assemble the sandwiches, spread a generous amount of muhammara on one side of **8 slices of sandwich bread**. Top 4 of the slices of bread with ¼ cup [12.5 g] of the dressed arugula and **1 sliced egg**. Season with **fine sea salt** and **ground black pepper**. Place the second slice of bread on top of each and serve.

continued

Refrigerate extra muhammara in an airtight container in the refrigerator for up to 1 week. It also freezes well for up to 1 month in a freezer-safe container. I've found that placing a piece of parchment paper on the surface of the muhammara or drizzling a layer of extra-virgin olive oil reduces the risk of dehydration and freezer burn.

I first fell in love with muhammara at one of Chef Reem Assil's restaurants, so much so that I quickly went on a path to learn how to make it. Muhammara is usually made with walnuts, but here the use of roasted peanuts with the roasted bell peppers makes the dip extra smoky. This recipe makes more muhammara than is needed for four sandwiches, but that is intentional (it makes about 2 cups [480 ml]). You don't want to make it every day, but rather make enough to last you a week so you can have it when-ever you want. This is a very unconventional and poten-tially controversial recommendation, but the muhammara is also very good with dosa and idlis.

THE COOK'S NOTES

- Roasted peanuts are readily available in stores, but if you like, you can roast your own. If you do, make a larger batch so you'll have some on hand for the future. Spread them on a rimmed baking sheet in a single layer and roast at 350°F [180°C] until they start to become fragrant and turn golden brown, stirring midway through cooking, about 30 minutes total. Cool completely before storing in an airtight container for up to 1 week in the refrigerator or 1 month in the freezer.

- Besides eggs, grilled or roasted slices of eggplant also go great in this sandwich.

276

Leek, Potato + Pancetta Soup

MAKES 2 SERVINGS AS A MEAL OR 4 SERVINGS AS A SIDE

Place **1 large starchy potato such as russet, peeled and diced into 1 in [2.5 cm] chunks**, in a medium saucepan and cover with enough water to submerge the potato completely by at least 1½ in [4 cm]. Stir in **1 tsp fine sea salt** and **¼ tsp baking soda**. Bring the water to a rolling boil over high heat. Lower the heat to a simmer and cook until the potato turns tender, about 10 minutes. Drain and transfer the potato chunks to a blender.

While the potatoes cook, prepare the leeks. In a medium stainless-steel skillet, warm **2 Tbsp extra-virgin olive oil** over medium heat. Add **2 large leeks, green parts trimmed and thinly sliced**, and **¼ tsp fine sea salt** and sauté until the leeks begin to fall apart and turn golden brown, 6 to 8 minutes. If the leeks begin to stick or burn at any time, add 1 tsp water, scraping the bottom of the pan.

Transfer the leeks to the blender with the potatoes. Add **3½ cups [830 ml] boiling water, 1 Tbsp apple cider vinegar, 4 garlic cloves, 2 Tbsp extra-virgin olive oil, 2 Tbsp white or yellow miso paste, ½ tsp ground turmeric or 15 strands of saffron, ½ tsp ground black pepper**, and **¼ tsp ground chipotle**. Using caution, blend carefully on high speed until smooth and silky. Strain the liquid through a fine mesh sieve held over a medium saucepan. Taste and season with fine sea salt.

Prepare the pancetta-leek garnish. Wipe the skillet clean (or rinse it if there are some stuck-on bits of leek) and heat over medium-low heat. Add **4 oz [115 g] diced pancetta** and cook until the fat completely renders and the pancetta starts to brown, 6 to 8 minutes. Add **1 tsp whole nigella** and **½ tsp ground black pepper** and sauté until fragrant, 30 to 45 seconds. Add **1 large leek, green part trimmed and thinly sliced**, and sauté until the leeks turn golden brown and begin to crisp, 6 to 8 minutes. The pancetta is salty, so taste first and season if you really need to with **fine sea salt**. Remove from the heat.

Divide the warm soup among four serving bowls and garnish with the pancetta-leek mixture. Serve immediately. Leftover soup can be stored in an airtight container in the refrigerator for up to 3 days.

continued

13. Bell Pepper, Eggplant, Potato, Tomato + Tomatillo

There are some soups that I prefer to eat in all their creamy jubilation, and potato soup is one of them. A creamy potato soup is a bit of a conundrum: We need the starch to thicken the soup, but we also don't want it to get to the point of becoming a gummy paste that could be used as a face mask. The Cook's Notes contain a few pointers to help you. The crispy leeks with the nigella and pancetta are a topping you might want to add to your repertoire because it goes well on several foods, including toast.

THE COOK'S NOTES

- After testing a variety of potatoes, I found that both russet and Yukon gold potatoes consistently produced the best results in terms of color, taste, and texture.

- I use leek greens, but some folks prefer not to. They are perfectly edible and very tasty. Slice them thinly, so they tenderize when cooking.

- Pancetta is salty, so be careful when salting the leek topping. Skip the pancetta if you don't consume it— and if you omit it, use 2 Tbsp extra-virgin olive oil to cook the leeks.

- Adding water and vinegar helps break the bonds that form between the starch granules inside the potatoes. The olive oil will also help coat the starch and prevent it from becoming gummy.

- If the soup thickens more than you prefer, gently whisk in 2 to 3 Tbsp boiling water at a time until it thins enough. Adjust the amount of salt, vinegar, and black pepper accordingly. Potatoes are notorious for sucking up flavors.

Veg-table

Coconut-Stuffed Baby Eggplants

MAKES 4 SERVINGS

Preheat the oven to 400°F [200°C]. Line a baking sheet or roasting pan with foil.

In the bottoms of **4 baby eggplants**, cut a cross pattern about two-thirds of the way through the eggplant. The cuts should not go all the way through the top, so the eggplant can hold its shape.

In a small mixing bowl, combine **¼ cup [20 g] grated fresh or frozen coconut (if frozen, thaw completely)**; **¼ cup [35 g] finely chopped cashews**; **1 shallot, finely minced**; **2 garlic cloves, minced or grated**; **2 Tbsp extra-virgin olive oil**; **1 Tbsp fresh lemon juice**; **1 tsp ground coriander**; **1 tsp ground cumin**; **1 tsp red pepper flakes such as Aleppo, Maras, or Urfa**; **1 tsp light brown sugar**; and **1 tsp fine sea salt**.

Using a fork or your hands, stuff the mixture into the crevices of the cut eggplants and place them on the prepared baking sheet. Rub the eggplants with **a little extra-virgin olive oil** and bake until the eggplants are tender and the flesh is easily pierced with a knife, 25 to 35 minutes, rotating the baking sheet halfway through cooking. Remove from the oven, garnish with **2 Tbsp torn fresh Thai basil**, and serve hot with **plain or Coconut Rice (page 295)**. Leftovers can be stored in an airtight container in the refrigerator for up to 3 days.

This is a simple and gratifying way to prepare eggplant. It performs exceptionally well as a casual weekday meal as well as at a more extravagant dinner party.

THE COOK'S NOTES

- If you're stuffing eggplant for the first time and are worried about the stuffing falling out, you can tie them with cooking twine after you stuff them. Cut and remove the twine before serving.

- Desiccated coconut doesn't work as nicely here as fresh. If you must use it, I recommend first hydrating it with 2 Tbsp boiling water.

Vegetarian-Stuffed Bell Peppers

Fill a medium saucepan with enough water to cover **2 or 3 small starchy potatoes, such as russets, scrubbed and cut into 1½ in [4 cm] chunks**. Stir in **1 tsp fine sea salt**. Bring the water to a rolling boil over medium-high heat. Turn down the heat to low and simmer until the potatoes are tender, 15 to 25 minutes. Drain the potatoes and place them in a large mixing bowl. Let rest until cool enough to handle, then remove and discard the peels. Smash the potatoes with a fork or, for a smoother texture, use a ricer.

Meanwhile, in a medium saucepan, bring **2 cups [480 ml] water, ½ cup [100 g] black lentils, and ¼ tsp fine sea salt** to a boil over high heat. Turn down the heat to low and cook until the lentils are tender but not falling apart, about 30 minutes. Drain the lentils and add to the large mixing bowl after the potatoes have been smashed.

Preheat the oven to 350°F [180°C]. Line a baking sheet with foil.

To the large bowl with the potatoes, fold in **1 shallot, minced; 2 Tbsp chopped cilantro; 1 green chilli such as serrano, jalapeño, or bird's eye, minced (optional); 1 Tbsp peeled and grated fresh ginger; 1 tsp ground Kashmiri chilli powder (or ¾ tsp smoked sweet paprika + ¼ tsp ground cayenne); ½ tsp ground turmeric; ½ tsp ground black pepper;** and **fine sea salt**.

In a small bowl, combine **¼ cup [15 g] panko, 2 Tbsp extra-virgin olive oil, ½ tsp ground black pepper**, and **fine sea salt**.

Trim the tops off **4 medium red, orange, or yellow bell peppers**, core, and discard the seeds. If the bell peppers can't stand on their own, trim the bottoms just enough so they can stand upright; it's fine if they're slightly crooked, this isn't the bell pepper Olympics. Stuff the bell peppers with the potato lentil mixture and top each with 1 Tbsp of the panko mixture. Brush the outsides of the bell peppers lightly with **a little extra-virgin olive oil**, place them on the prepared baking sheet, and bake for 30 to 40 minutes, rotating the pan halfway through baking, until the tops are crisp and golden brown, and the peppers are tender and can easily be pricked by a fork.

While the peppers bake, prepare the sauce. Add to a blender or food processor **½ cup [120 g] plain unsweetened Greek yogurt; ½ cup [60 g] crumbled feta; ¼ cup [35 g] raw or roasted pumpkin seeds (salted or unsalted); ½ tsp ground black pepper;**

2 Tbsp fresh lemon juice or apple cider vinegar; **1 bunch [115 g] cilantro, leaves and tender stems, cut in half**; and **1 green chilli such as serrano, jalapeño, or bird's eye, roughly chopped (optional)**. Pulse on high speed until combined; it's fine if the mixture is slightly textured. Taste and season with **fine sea salt**.

Remove the cooked peppers from the oven, place them on a small platter or plate, and serve with the sauce on the side. Leftovers can be stored in an airtight container in the refrigerator for up to 3 days.

My family loves to stuff vegetables with all sorts of fillings, such as ground meat and a medley of vegetables. Stuffed bell peppers with mashed potatoes are one of my favorite dishes. I've added lentils for some contrast to the texture of the smooth and creamy mash and topped them with crunchy bread crumbs. There's cool yogurt sauce to go with the stuffed peppers that is made with briny feta and pumpkin seeds.

THE COOK'S NOTES

- Keep in mind that boiling time for potatoes will vary by type and size.

- To convert this into an appetizer, stuff large jalapeños with the filling and bake them in the oven.

283

Breaded Harissa Eggplant + Lentil Salad

MAKES 4 SERVINGS

Preheat the oven to 350°F [180°C]. Line a rimmed baking sheet with foil.

In a small mixing bowl, combine **2 Tbsp extra-virgin olive oil**; **2 garlic cloves, grated**; **2 Tbsp harissa paste**; and **fine sea salt**.

In a second small mixing bowl, combine **¼ cup [15 g] panko**, **¼ cup [45 g] chia seeds**, **¼ cup [35 g] poppy seeds**, **2 Tbsp extra-virgin olive oil**, and **fine sea salt**.

Cut **2 large Japanese eggplants** in half lengthwise. Using a sharp paring knife, make a crosshatch pattern on the cut sides of the eggplant. The cuts should not go all the way through. Brush both sides with **a little extra-virgin olive oil**.

Lay the eggplant halves, cut side up, on the prepared baking sheet. Spoon and spread the harissa mixture over the cut side of each eggplant half and top with the panko mixture. Roast until the eggplant is tender and the tops are golden brown, 25 to 35 minutes. Remove from the oven and transfer to a serving plate.

While the eggplant cooks, prepare the lentil salad. In a medium saucepan, bring **4 cups [945 ml] water**, **1 cup [200 g] dried black lentils**, and **½ tsp fine sea salt** to a boil over high heat. Lower the heat to a gentle simmer and cook until the lentils are tender but not falling apart, about 30 minutes. Drain the lentils, transfer to a large mixing bowl, and let cool to room temperature.

To the mixing bowl, add **1 shallot, minced**; **¼ cup [30 g] crumbled feta or cotija**; **2 scallions, both white and pale green parts, thinly sliced**; **2 Tbsp chopped flat-leaf parsley**; **2 Tbsp chopped mint**; **2 Tbsp extra-virgin olive oil**; **2 Tbsp fresh lemon or lime juice**; **1 Tbsp date or maple syrup**; and **½ tsp ground black pepper**. Taste and season with **fine sea salt**.

Serve the warm eggplant with the lentil salad. This is best eaten the day it's made; leftovers can be refrigerated in an airtight container for up to 3 days. Reheat the eggplant in the oven at 300°F [150°C].

continued

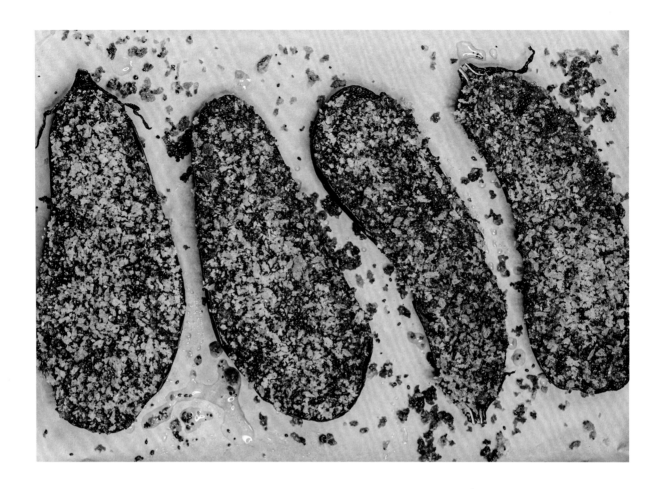

I am of the opinion that eggplant is one of the most glamorous vegetables in the looks department. Plus, the texture is gloriously satisfying, and the way its flesh shifts from spongy to silky smooth is simply marvelous. There's a very simple technique at play here: a few incisions on the eggplant to help the harissa permeate, and a topping of panko for texture. I've purposely paired the eggplant with this lentil salad, but if you'd like a different salad or an additional one, I highly recommend the Celery Herb Salad with Lime Vinaigrette (page 309).

THE COOK'S NOTES

- Eggplant shapes are individual variations on a theme. If you have trouble keeping your particular eggplants stable on their curved side, trim a tiny portion of the base to help it sit flat. Just a tiny portion!

- Scoring the eggplant with a crosshatch pattern helps the heat permeate the vegetable to the center and allows the harissa to go a little deeper.

- Before you use the harissa, take a tiny taste of it; I've noticed the intensity varies by brand. If the one you have isn't hot enough, you can double the quantity indicated here.

- The combination of bread crumbs, chia seeds, and poppy seeds gives the topping a nice crunch. It's better to spoon the topping than to use a spreading motion, or a lot of it will fall off.

286

Crispy Cauliflower with Tomatillo Salsa Verde

MAKES 4 TO 6 SERVINGS

Set two wire racks in the upper and lower thirds of the oven and preheat the oven to 425°F [220°C].

On a baking sheet pan, combine **8 oz [230 g] whole tomatillos, husked and rinsed**; **1 small white or yellow onion, diced**; **2 garlic cloves**; **1 jalapeño or serrano pepper, stemmed**; and **1 Tbsp extra-virgin olive oil** and spread in a single layer.

On a second baking sheet pan, toss **1 large head cauliflower, cut into bite-size florets**; **3 Tbsp extra-virgin olive oil**; **½ tsp ground black pepper**; and **fine sea salt** and spread in a single layer.

Place both baking sheets in the oven and roast for 25 to 30 minutes. About halfway through cooking, switch the pans between the racks and rotate the pans front to back. The pan with the tomatillos should be cooked until the vegetables start to char and the tomatillos burst and release their juices. The cauliflower should be roasted until golden brown and crisp.

Transfer the cooled roasted ingredients from the pan with the tomatillos to a blender or food processor. Add **½ cup [10 g] tightly packed cilantro, leaves and tender stems**, and **¼ cup [60 ml] fresh lime juice** and pulse on high speed for a few seconds to get a coarse mixture. Taste and season with **fine sea salt**. Sometimes depending on the tomatillos, a **tiny pinch of sugar** stirred in will help balance the taste. The salsa can be made a day ahead and stored in an airtight container in the refrigerator.

Once the cauliflower is cooked, place it in a large mixing bowl. The cauliflower can be roasted at least 6 hours ahead of time. Warm in a preheated oven at 300°F [150°C] before adding it to the bread crumb mixture (next step).

In a medium saucepan, warm **3 Tbsp extra-virgin olive oil** over medium-low heat. Add **4 garlic cloves, thinly sliced**, and sauté until golden brown, being careful not to overcook so it doesn't turn bitter. Remove the garlic with a fork or slotted spoon and transfer to the roasted cauliflower. To the same saucepan, add **½ cup [70 g] dried fine bread crumbs** or **1 cup [60 g] panko**, **¼ cup [35 g] dried sweet-tart cherries or sweetened dried cranberries**, **2 Tbsp chia seeds**, **2 Tbsp sesame seeds**, and **1 tsp dashi powder** (optional; see the Cook's Notes) and sauté until golden brown, 3 to 4 minutes.

continued

Remove from the heat and add to the cauliflower. Add the **zest of 1 lime** and toss to coat well. Taste and season with **fine sea salt**. Transfer to a serving plate, serve warm with the tomatillo salsa, and garnish with **2 Tbsp chopped cilantro**.

Leftovers can be stored in an airtight container in the refrigerator for up to 3 days. I sometimes eat leftovers tucked inside a sandwich or wrap.

This cauliflower dish does a fantastic job masquerading as a dressing or stuffing (which term you prefer is a war I'm happier staying out of), but it is also a wonderful way to enjoy roasted cauliflower. This roasted cauliflower is studded with sweet and tart cherries and a mix of bread crumbs and chia seeds and served alongside tomatillo salsa verde. To save on time, the cauliflower is roasted alongside the tomatillos and salsa ingredients.

THE COOK'S NOTES

- I'm leaving the dashi powder step optional. If the dashi you have on hand comes in the form of tiny granules, grind it down with a mortar and pestle before using. Dashi adds a lovely savory profile to the bread crumbs. A liquid stock won't work nicely here in place of dashi powder because the starch in the bread crumbs will bind the liquid and turn into something that resembles wet, clumpy sand.

- Tomatillos are naturally rich in pectin, which helps thicken the salsa as it stands.

Chaat-Style Loaded Twice-Baked Potatoes

MAKES 4 SERVINGS

Preheat the oven to 350°F [180°C]. Line a baking sheet with foil.

Scrub **2 large baking potatoes** and brush with **a little extra-virgin olive oil**. Place on the prepared baking sheet and bake until tender enough to be pierced with a knife all the way through, 60 to 75 minutes. Remove from the oven and let the potatoes rest for about 15 minutes until cool enough to handle. Leave the oven on.

Cut the potatoes in half lengthwise and scoop out most of the flesh using a spoon, leaving about ¼ in [6 mm] border of potato on the skin. Transfer the flesh to a large mixing bowl, making sure the skins are intact (it doesn't need to be perfect). Reserve the skins.

While the potatoes bake, prepare the date-tamarind chutney. In a small bowl, whisk together **¼ cup [60 ml] date syrup**, **2 Tbsp tamarind paste**, and **fine sea salt**. If the mixture is too thick, stir in **1 to 2 Tbsp boiling water**. The consistency should be thin and syrupy.

Mash the potatoes with a fork until there are no large clumps. Add **2 Tbsp extra-virgin olive oil**, **½ tsp ground black pepper**, and **fine sea salt** and fold gently to combine. Spoon into the reserved potato skins, place them on the baking sheet, and bake until the tops are light golden brown, about 15 minutes. Remove from the oven.

While the potatoes bake for the second time, in a small bowl, combine **2 shallots, minced**; **¼ cup [10 g] chopped cilantro**; **2 Tbsp chopped mint**; **1 green chilli such as serrano, jalapeño, or bird's eye, thinly sliced**; **2 Tbsp chaat masala, homemade (page 341) or store-bought**; and **ground kala namak** (see the Cook's Notes).

In a separate bowl, whisk **2 cups [480 g] plain, unsweetened full-fat Greek yogurt** and **fine sea salt**.

Top each hot potato with ½ cup [120 g] of the yogurt, followed by the shallot mixture. Drizzle with 1 Tbsp of the date-tamarind chutney and top with **2 to 3 generous Tbsp fine sev such as the "nylon" variety** (see the Cook's Notes). Serve immediately with **lime wedges** on the side, if desired. Leftover chutney can be kept in an airtight container in the refrigerator for up to 4 days.

continued

Potatoes play a very important role in Indian street food chaat-style dishes, so I figured, why not make a baked potato based on the same principles? I've limited the sauces to one—the date-tamarind chutney—and loaded the top with lots of fresh herbs and, of course, plenty of crunchy sev, a crispy noodle-like snack made from chickpea flour. If you want to add a fresh herb chutney, use the one listed in the Cook's Notes. For an even more loaded experience, make a batch of the Chana Masala (page 203; skip the pumpkin and rice) and serve it on top of the potatoes with the rest of the accoutrements.

THE COOK'S NOTES

- After the potatoes are baked, the flesh is scooped out carefully and mashed and seasoned, which makes every part of the potato flesh taste good.

- Often store-bought chaat masalas will contain kala namak (Indian black salt) already, so read the description and taste some before you decide to add any.

- Sev is a crispy noodle-looking snack made from chickpea flour. An alternative to sev is aloo bhujia, which looks the same and is made from potatoes. Both are available in Indian grocery stores and online. I prefer "nylon sev" because it is much finer and thinner than the others.

- I don't always serve this herb chutney because there are already a lot of herbs. But for those of you who want to use it, here's the recipe for my Quick Green Herb Chutney: Blend **1 cup [20 g] tightly packed cilantro**, **¼ cup [9 g] packed fresh mint with stems**, **1 green chilli such as serrano or jalapeño**, **2 Tbsp fresh lime juice**, **1 tsp toasted cumin**, and **1 tsp toasted coriander** until smooth and add enough water to make a thin chutney that's about 1 cup [240 ml]. Taste and season with **fine sea salt**.

292

Coriander, Cumin + Eggplant Noodles

MAKES 4 SERVINGS

In a large mixing bowl, combine **2 Tbsp low-sodium soy sauce**, **2 Tbsp Chinese black vinegar**, **1 tsp sugar**, and **¼ tsp fine sea salt**. Add **1 lb [455 g] globe eggplant, cut into ½ in [13 mm] chunks**. Toss to coat well. Let sit for 30 minutes.

Bring a large pot of salted water to a boil and cook **12 oz [340 g] dried chow mein stir-fry noodles** per the package instructions. Drain the cooked noodles in a colander, rinse under cold running water, and shake well to get rid of excess water. Leave in the colander and drizzle with **1 Tbsp toasted sesame oil** to prevent the noodles from sticking.

In a wok or large skillet, warm **2 Tbsp neutral oil with a high smoke point such as grapeseed** over high heat. Once the oil shimmers, add **1 large onion, thinly sliced**, and stir-fry until lightly browned, 6 to 8 minutes. Add **4 garlic cloves, thinly sliced**, and **one 2 in [5 cm] piece fresh ginger, peeled and cut into 1 in [2.5 cm] matchsticks**, and stir-fry until fragrant and the garlic has softened, 1 to 2 minutes. Add **1½ Tbsp ground cumin**, **1 Tbsp ground coriander**, and **1 tsp red pepper flakes** and stir-fry until fragrant, 30 to 45 seconds. Add **1 large bell pepper, cut into ½ in [13 mm] dice**, and stir-fry until softened and lightly brown, 3 to 4 minutes. Taste and season with **fine sea salt**. Remove from the heat and transfer to a large bowl.

Sprinkle **1 Tbsp cornstarch** over the marinated eggplant and toss to coat well.

Wipe the wok clean and warm **1 Tbsp neutral oil with a high smoke point such as grapeseed** over high heat until the oil shimmers. Add the eggplant and stir-fry until golden brown and tender, 4 to 5 minutes.

Return the onion and peppers to the wok. Fold to coat well. Fold in the cooked noodles and toss until warmed through. Drizzle with **2 Tbsp Chinese black vinegar**. Taste and season with **fine sea salt**. Garnish with **¼ cup [5 g] tightly packed fresh mint leaves, torn**. Serve immediately or warm. Leftovers can be stored in an airtight container in the refrigerator for up to 3 days.

continued

You're probably familiar with the immensely popular cumin-scented lamb stir-fries of Chinese and Mongolian cuisines. What I find fascinating is how this combination travels from southern parts of Asia all the way to India, where lamb is often seasoned with cumin (and coriander). Eggplant takes the place of lamb, and its spongy texture provides a satisfying alternative to the meat, absorbing the flavors from the vinegar and spices with ease. A sprinkling of fresh mint as the final touch is all that is needed.

THE COOK'S NOTES

- The spice quantities might seem like a lot, but it's what brings the flavor.

- Use hot red pepper flakes here.

- Chinese black vinegar is heavenly fragrant and can be found online or in Asian grocery stores. My favorite brand for this vinegar is Kong Yen.

294

Cashew + Bell Pepper Chicken with Coconut Rice

MAKES 4 SERVINGS

Coconut Rice

Rinse **1 cup [200 g] basmati rice** under running water until the runoff water is no longer cloudy. Transfer the rice to a bowl and add enough water to cover by 1 in [2.5 cm]; let soak for 30 minutes.

In a medium saucepan, combine **one 13½ oz [400 ml] can unsweetened, full-fat coconut milk**; **½ cup [120 ml] water**; **½ tsp fine sea salt**; and **¼ tsp ground green cardamom or 1 whole cardamom pod, cracked**. Bring to a boil over high heat. Drain the soaked rice and add it to the boiling mixture. Lower the heat to a simmer and cook, covered, until all the liquid is absorbed, 12 to 14 minutes. Remove from the heat and let sit, covered, for 5 minutes. Just before serving, fluff the rice with a fork.

In a small bowl, mix **1 Tbsp low-sodium soy sauce**; **1 Tbsp oyster sauce**; **1 Tbsp sesame oil, preferably toasted**; **1 Tbsp packed light brown sugar**; and **1 Tbsp ground black pepper**.

Heat a large wok or 12 in [30.5 cm] cast-iron skillet over high heat. Add **2 Tbsp neutral oil with a high smoke point such as grapeseed**. Add **1 medium white or yellow onion, diced into large chunks**, and stir-fry for 2 minutes. Add **4 garlic cloves, minced**; **4 dried whole red chillies**; **2 scallions, both white and pale green parts, cut into 1 in [2.5 cm] pieces**; and **1 tsp peeled and grated fresh ginger**. Stir-fry for about 1 minute, until fragrant. The garlic should not brown. Add **1 medium red bell pepper, cored and diced into large chunks**. Stir-fry for 1½ to 2 minutes, until they start to sear and slightly brown.

Add **1½ lb [680 g] boneless, skinless chicken breast or thigh meat, cut into 1 in [2.5 cm] chunks**. Stir-fry until the meat is browned on the outside and the internal temperature reaches 165°F [74°C], 3 to 5 minutes. Lower the heat to medium, pour the soy sauce mixture over the chicken, and toss to coat well.

In a small bowl, whisk **2 Tbsp water** with **1 Tbsp cornstarch** to form a smooth slurry. Stir into the chicken and cook until the liquid in the pan thickens and forms a sauce, 30 to 60 seconds. Remove from the heat. Taste and season with **fine sea salt**.

continued

Stir in **1 cup [140 g] roasted unsalted whole cashews**. Garnish with **2 Tbsp chopped chives** and immediately serve hot with the coconut rice. Leftovers can be stored in an airtight container in the refrigerator for up to 3 days.

Cashew chicken is one of my favorite dishes to order for takeout, because the copious quantity of crunchy cashews and succulent chunks of bell peppers coated in the savory sauce make me so happy. The coconut rice used here is a good option to pair with some of the other recipes in this book, such as Crispy Salmon with Green Curry Spinach (page 107) and Sweet + Sticky Brussels Sprouts (page 165), to name a few.

THE COOK'S NOTES

- For the hot, dried red chillies, I like to use Tianjin chillies, also known as Tien Tsin, but you can use any variety that you prefer. However, avoid smoked chillies such as chipotle in this recipe.

- If things get too crowded in the skillet, the bell peppers won't sear well. To avoid this, spread the vegetables out in an even layer as they cook.

Garlic Miso Steak with Roasted Bell Pepper Sauce

MAKES 4 SERVINGS

Preheat the oven to 425°F [220°C]. Line a baking sheet with foil.

Roasted Bell Pepper Sauce

In a large bowl, combine **1 large red bell pepper, cored and diced**; **1 pint [280 g] cherry or grape tomatoes**; and **2 Tbsp extra-virgin olive oil** and toss to coat. Spread in a single layer on the prepared baking sheet and roast until slightly charred, 18 to 24 minutes. Remove from the oven, let cool for 10 minutes, and transfer to a blender or food processor.

To the blender, add **2 garlic cloves**; **3 Tbsp malt vinegar, apple cider, or red wine vinegar**; **2 Tbsp extra-virgin olive oil**; **1 tsp smoked sweet paprika**; and **½ tsp ground chipotle** and pulse on high speed until combined but slightly chunky. Taste and season with **fine sea salt**. This sauce can be made ahead of time but tastes best when served at room temperature.

To prepare the steak, cut **1½ lb [680 g] flank or skirt steak crosswise into about 10 in [25 cm] lengths.** Rub the pieces with **2 Tbsp extra-virgin olive or grapeseed oil** and **1 Tbsp Worcestershire sauce** and let sit for 5 minutes. Lightly sprinkle the steak with **fine sea salt**; the steak will be brushed with a miso mixture after it is cooked, so don't oversalt.

Meanwhile, in a small mixing bowl, combine **2 Tbsp extra-virgin olive oil**; **1 Tbsp white or yellow miso paste**; **1 tsp fish sauce**; **2 garlic cloves, grated**; and **½ tsp ground black pepper** and set aside.

Heat a grill or grill pan over high heat. Brush the grates with **a little extra-virgin olive or grapeseed oil**. Place the steak on the hot grill and cook for 3 to 4 minutes per side for medium-rare (115°F [45°C] on an instant-read digital thermometer). Transfer to a plate.

Brush the steaks on both sides with the garlic miso mixture. Let sit for 10 minutes and then cut with a sharp knife against the grain into ¼ in [6 mm] slices. Serve warm with the roasted pepper sauce. Leftover steak and sauce can be stored separately in airtight containers and refrigerated for up to 3 days.

continued

It's no secret that I love roasted peppers, and I've roasted peppers quite a few times in this book. With any kind of sauce or dip, roasting the red pepper makes things smokier but also eliminates some of that rawness that I find tastes a bit awkward outside a salad. This is a vinegar-based red bell pepper sauce, and bolder-tasting vinegars like malt, apple, or red wine perform much better here than white wine or champagne vinegars. Skirt steaks are less expensive than other cuts of steak and quick to cook, but that also means you need to watch it carefully; read the Cook's Notes before proceeding.

THE COOK'S NOTES

- The recipe for the sauce makes a little more than you might need (but I could be wrong; you might end up loving it and finishing it in one sitting). If you do end up with leftovers, store in an airtight container in the refrigerator for up to 3 days. It goes great with the Bombay Potato Croquettes (page 266) and Vegetarian-Stuffed Bell Peppers (page 282) and is also great as a vegetable dip (Platters, Boards + Tricks, page 340).

- A note on the steak cooking temperature: Skirt steak is a thin cut of meat, and it cooks very quickly. As a time-saver this is great, but that also means you need to be extra careful to avoid overcooking it or it will become tough. Taking the steak off the heat at a lower temperature than you would with a thicker cut of meat (internal temperature 115°F [45°C]) helps avoid this problem, and the steak will continue to cook for a short while after it is removed from the pan.

300

Spaghetti with Roasted Tomato Miso Sauce

MAKES 4 SERVINGS

Preheat the oven to 400°F [200°C].

On a baking sheet, toss **3 lb [1.4 kg] cherry or grape tomatoes** with ¼ **cup [60 ml] extra-virgin olive oil** and roast until the tomatoes start to burst and turn a light golden brown, rotating the pan halfway through cooking, 25 to 30 minutes. Remove from the oven and transfer the tomatoes with the juices to a blender or food processor. Pulse on high speed into a smooth purée.

While the tomatoes roast, cook the pasta. Bring a large pot of salted water to a boil and cook **1 lb [455 g] dried spaghetti** until al dente, per the package instructions. Drain the cooked pasta and transfer to a large bowl.

In a medium saucepan, warm ¼ **cup [60 ml] extra-virgin olive oil** over medium heat. Add **4 garlic cloves, grated**, and **2 tsp red pepper flakes such as Aleppo, Maras, or Urfa** and swirl in the hot oil until fragrant and the oil starts to turn red, 30 to 45 seconds. Remove from the heat and whisk in ¼ **cup [40 g] white or yellow miso paste**. Stir in the puréed tomatoes until smooth and free of lumps. Return the saucepan to the stove, bring to a boil over high heat, then turn down the heat to low and simmer until the sauce thickens, 2 to 3 minutes. Taste and season with **fine sea salt**. Remove from the heat.

Fold the spaghetti with the tomato sauce. Garnish with ¼ **cup [15 g] grated Parmesan** and **torn fresh basil leaves**. Leftovers can be stored in an airtight container in the refrigerator for up to 4 days.

When I first created this pasta sauce for my newsletter, *The Flavor Files*, I was a little nervous. It deviates a lot from the norm, but it became one of the most popular recipes to date. Miso provides an instantaneous source of umami and balances out the taste of the sauce rather wonderfully. It also helps thicken the sauce.

THE COOK'S NOTES

- Roasting the tomatoes helps concentrate their flavors and makes them taste sweeter. The color of the sauce will vary depending on the color of the tomatoes you use, but that won't affect the final taste.

- You can opt for a hotter red pepper here, but I prefer a much gentler heat.

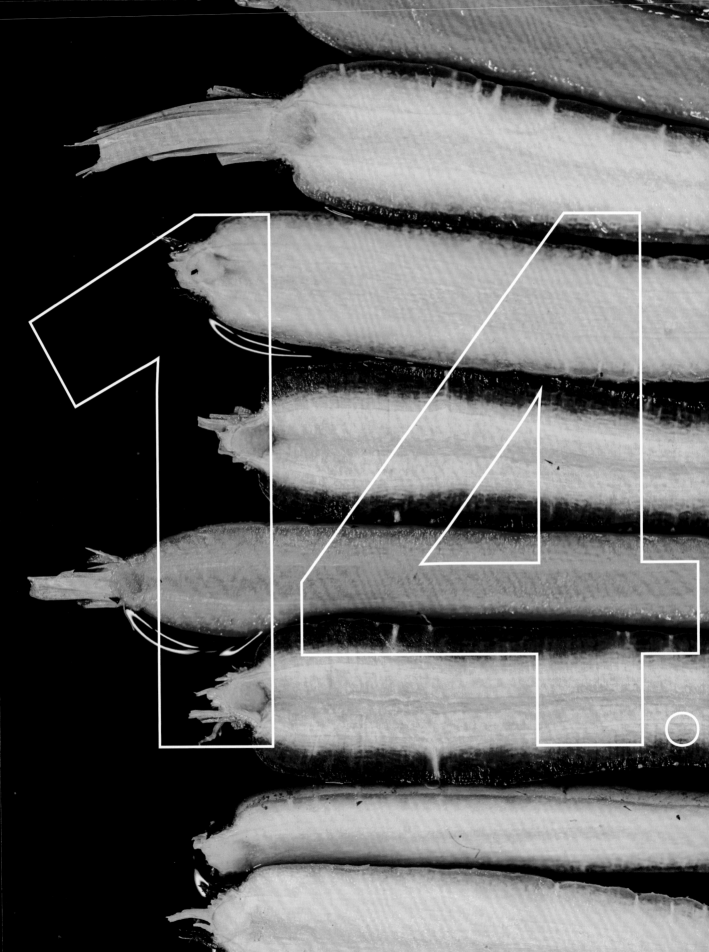

Carrots Celery Fennel + Parsnips

The Parsley Family

APIACEAE

Origins

CARROTS HAIL FROM AFGHANISTAN AND IRAN, CELERY AND FENNEL ORIGINATED IN THE MEDITERRANEAN, AND PARSNIPS ARE FROM EURASIA.

Carrots

Contrary to popular belief, younger carrots are not sweeter than older carrots. Research has shown that older carrots are richer in sugar and carotenes, making them a great choice for cooking. There are so many colorful varieties when it comes to carrots—you can feast on a rainbow with these vegetables.

Celery

Without celery, stocks lack personality, and stuffing just doesn't taste as nice. Both the greens and stalks are edible. This is also my dog Paddington's favorite snack and I grow some in my garden just for him.

Fennel

I once worked with an Italian who came to work with a small box filled with thinly shaved slices of fresh fennel that would be consumed after every meal or even as a snack. The sweet licorice-like smell of the fennel would take over the room as soon as the box was cracked open, and it was wonderful. Both the bulb and the fronds are edible and can be eaten raw as well as cooked. In California, fennel grows wild and will show up in the most unexpected places with its bright yellow flowers in spring and is quite a spectacular sight to behold.

Parsnips

Parsnips might look like carrots, but they are sweeter and contain more sugar. Parsnips are creamy white in color, and they can be eaten raw but taste better cooked. If you do eat them raw, shave thinly for a salad.

Storage

Carrots can be stored in an airtight plastic bag in the refrigerator or in a bowl of filtered water that must be changed daily. Cut the tops of carrots before storing them, and the greens can be wrapped in paper and kept in a ziptop bag. Celery and fennel can be stored wrapped in paper towels and kept inside a ziptop bag in the refrigerator. To store fennel, trim the frondy top of the bulb and leave about 2 in [5 cm] of the stalk at the top. Parsnips can be stored loosely wrapped in dry paper towels in an airtight container in the refrigerator. Small- to medium-size parsnips can also be cut into smaller pieces, blanched in boiling water for 3 to 4 minutes, cooled in an ice-water bath, patted dry with paper towels, and frozen on a baking sheet. Transfer the frozen pieces to an airtight container or bag and freeze for up to 6 months.

Cooking Tips

- Carrot greens can be used to make pesto as well as various types of herb blends. Fresh celery leaves and tender fronds from fennel are great as herbs in dishes.

- When preparing fennel for fresh preparations, trim and discard any tough ends. Fennel tastes best when cut or shaved thinly. Once cut, immerse the slices in a bowl of ice-cold water and drain before use. This helps maintain the crisp texture of the vegetable.

- Skip peeling young carrots, as their skin is too thin. Simply rinse well to remove and dislodge any dirt. Older carrots can be peeled, but in general, it is more advantageous to scrape the skin off with a knife than with a vegetable peeler to reduce waste.

- Older carrots can turn tough over time, but they can be used to make stocks and soups.

- Young or new carrots refer to the ones that are small and not the baby carrots that are prepared by shaving down larger carrots to create rounded tips.

- Carrots and parsnips are high in pectin, the structural carbohydrate that provides firmness. When making carrot or parsnip soup, add a small amount of baking soda, which helps solubilize the pectin, resulting in a much creamier soup. The baking soda also helps accelerate caramelization and the Maillard reaction of the sugars and liquid aminos in the vegetables (Carrot, Apple + Harissa Soup, page 312).

Sometimes larger parsnips can benefit from having their inner tough core removed. Admittedly, I don't do this often unless the parsnip is huge.

Both carrots and parsnips benefit from being cooked in a fat like

- Both carrots and parsnips are great as purées. I prefer to roast them first (carrots at 400°F [200°C] and parsnips at 425°F [220°C]), then blend them until smooth, adding water as needed. Roasting helps improve the flavor and makes them taste sweeter, unlike boiling.

Celery Herb Salad with Lime Vinaigrette

MAKES 4 SERVINGS

Lime Vinaigrette

In a small mixing bowl, whisk together **3 Tbsp fresh lime juice**; **2 Tbsp walnut or grapeseed oil**; **1 Tbsp honey, agave, or maple syrup**; **½ tsp toasted cumin seeds, lightly cracked**; **½ tsp whole celery seed**; **½ tsp ground black pepper**; and **fine sea salt**. Stir in **2 Tbsp chopped fresh chives**, **2 Tbsp chopped fresh dill**, **2 Tbsp chopped fresh mint**, and **2 Tbsp chopped fresh tarragon** and let sit for 15 minutes.

In a large mixing bowl, combine **6 large stalks celery and any tender leaves (about 1¼ lb [570 g]), thinly sliced**, and **½ cup [90 g] dried sweet-tart cherries or dried sweetened cranberries**.

When ready to serve, pour the vinaigrette over the celery in the large bowl and toss to coat well. Taste and season with **fine sea salt** if needed. Garnish with **¼ cup [15 g] shredded Parmesan cheese** and serve immediately. Leftover salad can be stored in an airtight container in the refrigerator for up to 1 day.

Celery is so strongly associated with stocks that it doesn't get enough attention in the kitchen outside soups or in those veggie dip trays from the grocery store. There's one more reason, and it might have to do with genetics: an aversion to celery's unique aroma and texture. The Cook's Notes has some suggestions that might be useful. This is a refreshing celery salad that uses a lot of herbs and a fragrant lime juice vinaigrette with pops of sweet and tart goodness from dried cherries.

THE COOK'S NOTES

- For people who aren't too keen on celery, I recommend cutting the celery stalks finely or even shaving them with a peeler. You can soak the cut slices in a bowl of cold water, drain, and then use it in the salad. These methods help reduce the intensity of the celery flavor.

- In a pinch, lemon juice will also work here, but the fragrance of lime is much better in this dish.

Hasselback Parsnips with Pistachio Pesto

MAKES 4 SERVINGS

Preheat the oven to 425°F [220°C]. Line a baking sheet with parchment paper.

Trim and peel **4 large parsnips** and slice lengthwise, then make crosswise cuts that are ¼ in [6 mm] apart; the cuts should not go all the way through (see facing page).

Bring a large saucepan of salted water to a boil over high heat. Carefully drop the parsnips in the water and poach for 1 minute. Remove the parsnips carefully with kitchen tongs or a slotted spoon and transfer them to the prepared baking sheet, flat side down.

In a small bowl, combine **¼ cup [60 ml] extra-virgin olive oil** and **½ tsp ground black pepper** and brush the mixture all over and deep into the grooves of the parsnips. Season with **fine sea salt**. Roast until crisp around the edges and tender enough to be easily pierced with a knife, 30 to 40 minutes. Remove from the oven.

Pistachio Pesto

While the parsnips roast, prepare the pesto. To the bowl of a food processor, add **4 cups [80 g] tightly packed fresh basil, leaves and tender stems**; **½ cup [70 g] unsalted raw pistachios**; **¼ cup [60 ml] extra-virgin olive oil**; **¼ cup [15 g] grated Parmesan**; and **2 garlic cloves** and pulse for a few seconds until a coarse mixture is obtained. Taste and season with **fine sea salt**. Add more **olive oil** as needed to make the sauce a loose consistency.

Spoon 1 Tbsp of the pesto over each of the hot parsnips and serve the remainder on the side. Season with more **fine sea salt** as necessary. Leftovers can be stored in an airtight container in the refrigerator for up to 3 days.

Hasselbacking is the equivalent of a costume party for vegetables—it gives tough vegetables like potatoes, butternut squash, and even parsnips a makeover. The accordion-like fan appearance helps heat penetrate better and creates a crisp texture on the "fans" while the center turns tender.

THE COOK'S NOTES

- When hasselbacking, make sure your cuts go deep but never all the way through. The center needs to hold the vegetable together. If you accidentally do cut the parsnip in half, place the halves cut side adjacent to each other.

- Season the crevices between each cut. A pastry brush will serve you well.

Carrot, Apple + Harissa Soup

Preheat the oven to 400°F [200°C].

In a large bowl, combine **1½ lb [680 g] carrots, trimmed and peeled, sliced diagonally into ½ in [13 mm] thick slices (about 4 cups)**; **1 large (7¾ oz [220 g]) Granny Smith apple, peeled, cored, and sliced into 1 in [2.5 cm] wedges**; **1 Tbsp extra-virgin olive oil**; **½ tsp fine sea salt**; and **¼ tsp baking soda**. Spread on a baking sheet and roast until golden brown and tender (a knife should be able to pierce through the carrots and apples with minimal resistance), about 25 minutes.

While the carrots and apples are roasting, prepare the topping for the soup. In a small saucepan or skillet, warm **2 Tbsp extra-virgin olive oil** over medium-high heat. Add **2 Tbsp sunflower seeds or pine nuts, ½ tsp celery seeds or caraway seeds**, and **flaky salt** and fry until the sunflower seeds turn golden brown, 1½ to 2 minutes. Remove from the heat and place in a heatproof bowl. Add **½ tsp smoked sweet paprika** and stir to coat.

Transfer the roasted carrots and apples to a blender or food processor. Add **3½ cups [830 ml] water, 2 Tbsp fresh lemon juice, 1 Tbsp peeled minced fresh ginger**, and **1 tsp harissa paste**.

Blend until smooth and velvety. Taste and season with **fine sea salt**. The final volume should be about 6 cups [1.4 L]; if needed, add more water.

Pour the blended soup into a medium saucepan and warm over medium heat. When ready to serve, top each bowl with a sprinkle of the sunflower seed mixture and serve hot or warm. Leftovers can be stored in an airtight container in the refrigerator for up to 3 days.

Harissa is a hot and spicy aromatic condiment made from chillies that originated in North Africa. In this carrot soup, it walks in like Mariah Carey at a concert and shakes everyone up with a spectacular reverberation of flavor that can be felt all the way through. The roasted apples round out the flavor of the carrots with their tart and sweet notes. Toasted buttered slices of sourdough are a worthy sidekick here.

THE COOK'S NOTES

- If you like the fragrance of roses, try rose harissa here. It gives a wonderful mild aroma of roses to this soup. New York Shuk makes an excellent rose harissa seasoning.

- Roasting the apples and carrots with baking soda helps soften them and bring out their bittersweet flavors by accelerating caramelization and the Maillard reaction.

312

Mustard Chicken, Fennel + New Potatoes

MAKES 4 SERVINGS

Preheat the oven to 200°F [95°C].

In a large mixing bowl, combine **¼ cup [60 g] whole-grain Dijon mustard**, **1 Tbsp light-brown sugar**, and **1 tsp low-sodium soy sauce**.

Pat dry on both sides with clean paper towels **4 large (total weight about 1½ lb [680 g]) bone-in, skin-on chicken thighs** (see the Cook's Notes). Season on both sides with **fine sea salt** and **ground black pepper**.

In a large stainless-steel skillet, warm **3 Tbsp extra-virgin olive oil** over medium heat. When the oil is hot, add the chicken, skin side down, and cook until browned on both sides, 5 to 8 minutes—no need to cook the chicken all the way through at this stage. Transfer the chicken to the bowl with the mustard marinade and gently turn to coat well. Reserve any fat remaining in the skillet.

In the same skillet, melt **2 Tbsp unsalted butter** over medium-low heat. Cook until the milk solids turn light brown and the liquid stops crackling, then add **2 Tbsp extra-virgin olive oil**. When the oil is hot, add **1 lb [455 g] baby potatoes, cut in half lengthwise**. Shake the pan to coat the potatoes in the fat, cover, and cook until the potatoes are golden brown and crisp, 10 to 12 minutes. Stir the potatoes occasionally during cooking (be careful of splattering) and lower the heat as necessary to prevent burning. If things start to burn, add 2 to 3 Tbsp water. Remove the potatoes with a slotted spoon and transfer to a small baking dish. Sprinkle with **fine sea salt**. Keep warm in the oven.

Clean the skillet and return to the stove over medium-high heat, adding **a drizzle of olive oil**. Add the marinated chicken thighs, skin side down, with all the marinade and cook on both sides, flipping them as they turn golden brown. Stir in **1 cup [240 ml] water or low-sodium chicken stock**. Cover and simmer over low heat until the chicken is cooked through and the internal temperature reaches 165°F [74°C], 10 to 12 minutes.

Uncover the pan and fold in **1 cup [120 g] frozen peas** (no need to thaw) and **1 large fennel bulb, trimmed and thinly sliced lengthwise**. Cover and cook until the peas and fennel are tender, 2 to 3 minutes. Remove from the heat. Top with the crispy fried potatoes and garnish with **2 Tbsp chopped chives** and **2 Tbsp chopped tender fennel fronds**. Serve hot or warm.

continued

This recipe is based on a Dijon mustard chicken recipe that first appeared in *A Bird in the Hand*, one of many great cookbooks by my dear friend Diana Henry. In this version, the chicken is first seared on the skin side just to brown and develop the flavors, then tossed in the mustard marinade. At the final stage, aromatic slices of fennel are added to imbue the dish with their fragrance. This is an easy weekday meal that will please many.

THE COOK'S NOTES

- You can use chicken thighs with the bone left in, with or without skin, depending on your preference. I like the bone and skin because of the extra flavor and texture they provide.

- The marinade on the chicken can burn easily, so watch closely when cooking on the stove.

- I usually like to use my cast-iron skillet whenever possible. You can use it here, but I find it makes a bigger mess.

316

Carrot Frittata

MAKES 4 SERVINGS

Preheat the oven to 400°F [200°C].

In a 12 in [30.5 cm] cast-iron skillet, melt **2 Tbsp unsalted butter** over medium-high heat. Once the butter stops crackling and the water evaporates, add **2 Tbsp extra-virgin olive oil** and swirl to combine. Add **4 garlic cloves, thinly sliced**, and **1 tsp ground coriander** and cook until fragrant, 30 to 45 seconds. Stir in **2 medium carrots, shredded**, and a tiny pinch of **fine sea salt**. Cover with a lid and cook, stirring occasionally, until the carrots are tender, 3 to 4 minutes.

While the carrots cook, prepare the eggs. In a large bowl, whisk together **8 large eggs**, **¼ cup [60 ml] whole milk**, **1 tsp fine sea salt**, **½ tsp baking soda**, and **½ tsp ground turmeric** until smooth and there are no lumps of baking soda or turmeric visible. (If you're having a hard time incorporating them with a whisk, you can give a quick blitz with an immersion blender.) Fold in **1 cup [20 g] packed baby arugula**; **3 scallions, both white and green parts, thinly sliced**; and **¼ cup [2.5 g] chopped fresh dill**.

Using a spoon or silicone spatula, spread the carrots in a thin layer on the surface of the skillet. Pour the egg mixture over the carrots and cook until the sides just begin to firm up, 1 to 3 minutes. Transfer the skillet to the oven and cook until the eggs start to set and firm up and the edges turn slightly golden brown, 3 to 4 minutes.

Remove from the oven, sprinkle **2 Tbsp grated Parmesan** over, return to the oven, and cook until the cheese melts. Remove and serve warm with the Cashew Green Chutney (page 93). Any leftovers can be wrapped with plastic wrap and stored in an airtight container in the refrigerator for up to 2 days.

Frittatas are more than just a way to use up leftovers—they make elegant centerpieces for brunch, lunch, and dinner. They're also a one-pan dish that doesn't really need anything else to accompany it, and I love that feature above all. You will need to cook the carrots a little before they meet the eggs to help tenderize the vegetable.

THE COOK'S NOTES

- Caraway is a spice that works great with carrots. Add 1 tsp crushed caraway seeds and sauté alongside the garlic and coriander.

- The baking soda helps the eggs rise during cooking and they hold air much better on baking.

Nopalitos

The Cactus Family
CACTACEAE

Origins
NOPALITOS HAIL FROM MEXICO AND CENTRAL AMERICA.

Nopalitos, Cactus, or Cactus Paddles

In Spanish, *nopales* refer to the cactus stem on the plant and *nopalitos* refer to the pads once they are cut and removed from the plant. Nopalitos are available in their native prickly state or pre-cleaned in Mexican and Latin American markets and stores, but I also find them in my grocery store. You can grow your own cactus indoors or outdoors depending on where you live. The plant prefers warm, sunny locations and will quickly reward you with multiple paddles, flowers adorned with papery petals, and fruit. The pads should be a fresh, bright ver- dant green, the skin on the pad should be tight and not wrinkled, and when touched gently it should feel slightly soft yet firm.

Storage

Wrap in dry paper towels and store in a plastic bag in the crisper drawer of the refrigerator for up to 4 days.

Cooking Tips

- **Preparing nopalitos:** Most of what I've learned about nopalitos and Mexican cuisine comes from author and teacher extraordinaire Pati Jinich. Preparing nopalitos isn't tricky, but it requires a little patience. The cactus paddles need to be picked of their spines and I always think of it as picking out scales or tiny bones from fresh fish.

Be careful when handling nopalitos, as those tiny thorns might seem harmless but can easily prick your skin. I usually wear a clean pair of garden gloves to handle them because they're thicker than regular gloves; if you feel comfortable, you can also handle them with folded kitchen towels or newspaper (that's how I usually handle cacti that I plant in my garden). Wash the nopalitos under running water to remove dirt. Use a vegetable peeler or a knife to get rid of all the thorns. Rinse the nopalitos again and pat dry with a clean kitchen towel, then trim and discard ½ in [13 mm] from the bottom of the paddle and ¼ in [6 mm] from the sides. Many grocery stores sell nopalitos already cleaned but double-check them carefully like a hawk before you use them.

- **Cooking nopalitos:** Nopalitos release a viscous liquid during boiling or stewing, just like okra (see page 250). This thick mucilage is made up of carbohydrates. If you want to get rid of it, a pinch of baking soda added to a pot of boiling salted water will help reduce some of its thickness.

- Like spinach, fresh nopalitos are rich in oxalic acid, and they will etch your teeth. Cooking the pads breaks down the oxalic acid, eliminating this unpleasant sensation.

- Grilling and searing are some other ways to cook nopalitos. Whole nopalitos can be brushed with oil and grilled until they develop nice sear marks and turn tender. Slice and use as needed.

320

Nopalito Bean Salad

MAKES 4 SERVINGS

Bring a medium pot of salted water to a rolling boil over high heat. Set up an ice bath in a large bowl.

To the boiling water, add **2 or 3 large nopalitos (total weight about 13½ oz [385 g]), cleaned and prepared (see page 320), chopped into 1 in [2.5 cm] pieces**. Cook until dull green and very tender, 15 to 20 minutes. Transfer the nopalitos with a slotted spoon to the ice bath. Let cool for 1 minute. Drain and rinse the nopalitos with running water. Shake to get rid of any excess water, blot dry with a kitchen towel, and transfer to a large mixing bowl.

To the bowl, add one **15 oz [425 g] can black or kidney beans, drained and rinsed**; **1 pint [280 g] cherry or grape tomatoes, sliced in half lengthwise**; **2 shallots, minced**; **4 scallions, both white and green parts, thinly sliced**; **¼ cup [30 g] crumbled feta or cotija**; **2 Tbsp chopped cilantro**; and **2 Tbsp roasted salted pumpkin seeds**. Toss to combine.

In a small bowl, whisk together **¼ cup [60 ml] neutral oil with a high smoke point such as grapeseed**; **2 Tbsp fresh lime juice**; **½ tsp ground black pepper**; **½ tsp red pepper flakes such as Aleppo, Urfa, or Maras**; **½ tsp ground toasted cumin** (see the Cook's Notes); and **fine sea salt**. Pour the dressing over the salad. Taste, add more **fine sea salt** as needed, and serve immediately.

Nopalitos love briny flavors; add a salty cheese like feta or cotija, and they will dance, as they do in this bean salad. Another way to eat this salad is piled in a taco shell and topped with Guajillo Chilli Salsa (page 138).

THE COOK'S NOTES

- Toast ¼ cup [30 g] whole cumin seeds in a dry stainless-steel skillet over medium heat until they are fragrant and turn light brown, 30 to 45 seconds. Transfer to a plate to cool completely before grinding to a fine powder. Use as needed and store leftovers in an airtight jar away from light in a cool, dark spot for up to 6 months.

- If you want to skip the feta or cotija, use brined olives instead for their salty taste.

- Nopalitos behave like cut cucumbers: They release a lot of water on standing due to osmosis. Put this salad together as soon as you're ready to serve.

322

Grilled Nopalito Sesame Salad

MAKES 4 SERVINGS AS A SIDE

Brush **2 or 3 large nopalitos (total weight about 13½ oz [385 g]), cleaned and prepared (see page 320),** with **1 to 2 Tbsp neutral oil with a high smoke point such as grapeseed**. Season with **a little fine sea salt** on each side.

Heat a grill or a grill pan over medium-high heat. Brush the grates with **a little grapeseed oil**.

Cook the nopalitos on each side until they start to change color and develop char marks, 8 to 10 minutes per side. Transfer to a cutting board and cut them crosswise into ¼ in [6 mm] strips and place them in a large mixing bowl.

Heat a small dry stainless-steel skillet over medium-high heat. The pan is hot when a drop of water sizzles and evaporates immediately. Add **1 Tbsp white sesame seeds** and **1 Tbsp black sesame seeds** and toast, stirring constantly, until the white sesame seeds start to turn a light golden brown (it's easier to rely on the white sesame seeds than the black sesame seeds for the change in color) and are fragrant, 30 to 45 seconds. Transfer the seeds to a mortar and pestle and crush gently.

Prepare the vinaigrette. In a small bowl, combine **3 Tbsp Chinese black vinegar or rice wine vinegar, 1 Tbsp toasted sesame oil, 2 Tbsp mirin, 1 tsp low-sodium soy sauce,** and **½ tsp ground black pepper**. Add the toasted sesame seeds and the vinaigrette to the nopalitos in the bowl. Add **1½ cups [80 g] packed watercress** and **1 shallot, minced**. Toss to coat well. Taste and season with **fine sea salt**.

Transfer to a serving dish and serve immediately. This salad is best eaten right away because the nopalitos will continue to release liquid as they sit in the vinaigrette and the watercress will turn limp.

In need of a new side salad for your next pizza, barbecue, or burger party? Look no further. This grilled cactus pad salad is quick to put together and light enough to complement the heavier fare of the main course. This is meant to be served as a small side salad, but sometimes when I want to make this a bit more substantial I'll fold in grilled or roasted potatoes to bulk it up.

THE COOK'S NOTES

• Baby arugula is a good alternative to watercress; it carries a similar peppery bite.

• Watercress is tender and wilts easily. Once the salad is put together, eat it soon to avoid this.

325

Nopalito + Chickpea Coconut Curry

MAKES 4 SERVINGS

In a medium saucepan, warm **2 Tbsp virgin coconut oil** over medium-high heat. Add **1 large yellow or white onion, chopped**, and sauté until it turns translucent and just starts to brown, 4 to 5 minutes. Add **4 garlic cloves, grated**, and sauté until fragrant, about 45 seconds. Add **1 tsp whole cumin seeds, ½ tsp ground turmeric, ¾ tsp smoked paprika powder**, and **¼ tsp ground cayenne**. Cook until fragrant, 30 to 45 seconds.

Lower the heat to medium and add **¼ cup [60 g] tomato paste**. Sauté for 3 to 4 minutes, until the paste starts to darken, scraping the bottom of the saucepan and adding 1 to 2 Tbsp water if needed to prevent scorching. Add **2 or 3 large nopalitos (total weight about 13½ oz [385 g]), cleaned and prepared (see page 320) and cut into ¼ in [6 mm] chunks**, and one **14 oz [400 g] can chickpeas**, rinsed and drained.

Stir in **1 cup [240 ml] plain, unsweetened full-fat coconut milk**. Turn down the heat to low and simmer, covered, stirring occasionally, until the nopalitos are soft and turn dull green, 10 to 13 minutes. Taste and season with **fine sea salt**.

Garnish with **2 Tbsp chopped cilantro** and serve hot or warm with **flatbread** or **rice**.

Cactus in a curry? Believe me, it's excellent. In this curry inspired by Goan coastal flavors, the cactus provides a lovely tender and juicy counterpart to the chickpeas. This is a hearty meal, one that needs no accompaniment other than a carbohydrate. While I'm all about rice being the number one starch to accompany most dishes, this is one place where I prefer flatbreads like naan, parathas, or rotis.

THE COOK'S NOTES

- I prefer to use virgin or unrefined coconut oil for a more pronounced dose of coconut aroma. Refined coconut oil is filtered and carries little to no fragrance.

- Tomato paste is the shortcut. It skips the extra time that would otherwise be needed to cook down fresh tomatoes.

Mixtape

Masala Veggie Burgers

Preheat the oven to 200°F [95°C]. Line a baking sheet with foil.

Toss one **15 oz [425 g] can black beans, rinsed and drained**, with **1 Tbsp extra-virgin olive oil**. Spread in a single layer on the prepared baking sheet and cook until most of the liquid evaporates and the beans are dry but not crisp, 30 to 45 minutes. Transfer the beans to the bowl of a food processor.

Meanwhile, in a cast-iron skillet or large saucepan, warm **2 Tbsp extra-virgin olive oil** over medium heat. Add **8¾ oz [250 g] sweet potato, peeled and coarsely grated; 2¼ oz [65 g] carrot, peeled and coarsely grated; and 1 medium celery stalk (about 1 oz [30 g]), chopped**. Sauté until the vegetables release most of their water and the water evaporates, 6 to 8 minutes. Remove from the heat and let cool.

Transfer the mixture to the food processor along with **¾ cup [150 g] day-old sticky rice, chilled or at room temperature; ½ cup [20 g] chopped cilantro; ¼ cup [60 g] tomato paste; 1 Tbsp onion powder; 2 tsp garlic powder; 1 tsp low-sodium soy sauce; 1 tsp nutritional yeast (optional); 1 tsp garam masala, homemade (page 341) or store-bought; 1 tsp fine sea salt; ½ tsp ground turmeric; and ¼ tsp ground cayenne**. Pulse on high speed until the mixture is coarsely ground; it should not be a smooth paste, or it will be very difficult to bind and cook. Transfer to a large mixing bowl, cover with a lid or cling film, and let sit for at least 30 minutes, but no longer than 1 hour, at room temperature.

Line a baking sheet with parchment paper. Divide the mixture into eight equal parts by weight. Grease your hands with **a little olive oil** and shape each part into a rough 3 to 4 in [7.5 to 10 cm] wide patty, about ¾ in [2 cm] thick (each burger will weigh about 3½ oz [100 g]). Lay the patties on the prepared baking sheet and freeze them, uncovered, for at least 30 minutes until they harden. Wrap the individual burger patties in parchment paper and store them flat in an airtight freezer-safe container or ziptop bag. The burgers can be stored for up to 1 month in the freezer.

continued

When ready to cook, in a cast-iron or stainless-steel skillet, warm **2 Tbsp neutral oil such as grapeseed** over medium heat. Add the frozen burgers and fry for 2 to 3 minutes per side, until the burgers start to crisp, turn golden brown, and release themselves from the skillet. Serve the burger inside your favorite **bun or lettuce** with your **favorite fixings**.

Imagine trying to make everyone in a class or a family at Thanksgiving get along. It's going to require a bit of effort, planning, and patience. The problem here is there are too many different personalities coming together in one spot: That's a veggie burger for you. Veggie burgers are surprisingly tricky to make because they include all sorts of different vegetables with different types and varying amounts of carbohydrates, proteins, and fats, and a lot of moisture. To coax them all together, I use a piece of advice from the *New York Times* columnist and cookbook author Melissa Clark: Dry the out vegetables. I then bind them with starch from cooked rice (cornstarch and flour leave a gummy texture). By the way, I've intentionally left out fixings and how to assemble a burger—for most people that's a very personal experience, so build yours as you like it.

THE COOK'S NOTES

- Because this recipe relies on ratios to be successful, the weights are listed when necessary as vegetables can be unwieldy in their sizes.

- The more water in the burger mixture, the trickier it becomes for the mixture to hold. Make sure you get rid of as much water as you can from the beans and the vegetables during cooking.

- Using tomato paste, onion powder, and garlic powder helps provide flavor without contributing water.

- The starch from the beans, sweet potatoes, and carrots will act as a binding agent.

- By comparison, sticky rice does a much more efficient job of holding the burger together than long-grain rice like basmati or jasmine.

- The chia seeds help absorb any excess moisture.

- In the worst-case scenario that you're still having problems binding the mixture, fold in ½ cup [70 g] dried bread crumbs.

333

Nigella-Spiced Vegetable Medley

MAKES 6 TO 8 SERVINGS

Preheat the oven to 350°F [180°C].

To a large bowl, add **2 large russet potatoes, peeled and cut into ¼ in [6 mm] matchsticks**; **1 lb [455 g] broccoli florets, cut into bite-size pieces**; **12 oz [340 g] cauliflower florets, cut into bite-size pieces**; **2 medium carrots, trimmed, peeled, and cut into ¼ in [6 mm] matchsticks**; **1 medium green bell pepper, cored and cut into ¼ in [6 mm] strips**; and **3 shallots, cut in half.**

In a small bowl, combine **¼ cup [60 ml] extra-virgin olive oil or ghee**; **1 Tbsp peeled and grated fresh ginger**; **2 tsp ground coriander**; **1 tsp red pepper flakes such as Aleppo, Maras, or Urfa**; **1 tsp whole cumin seeds**; **1 tsp nigella seeds**; **½ tsp ground black pepper**; and **1 tsp fine sea salt**. Add the spice mixture to the vegetables and toss to thoroughly coat.

Spread the vegetables on a rimmed baking sheet or roasting pan. Wrap the baking sheet with one or two layers of foil and fold the ends to form a tight seal. Roast for 45 minutes. Remove from the oven, unwrap, and check whether the vegetables are tender; a knife or skewer should pass through with ease. If they need more time, rewrap and cook for an additional 10 minutes and check for doneness.

Once the vegetables are cooked, transfer them to a serving plate. Drizzle with **2 Tbsp fresh lemon or lime juice** and garnish with **2 Tbsp torn fresh mint**; **2 Tbsp cilantro**; and **1 fresh chilli such as jalapeño, serrano, or bird's eye, thinly sliced.**

Serve immediately with **rice** or **flatbread**. This is also quite lovely as an accompaniment to a warm silky-smooth soup like the Leek, Potato + Pancetta Soup (page 277). Store leftovers in an airtight container in the refrigerator for up to 3 days.

continued

When I was growing up in India, ovens weren't a standard kitchen appliance, so you can only imagine my joy when we brought home our first oven. Our new oven was a small countertop model, and I learned not only how to bake cakes in it but also how to roast vegetables. On weekends when my parents let me make dinner, this dish was always in rotation. Baked spiced vegetables are seasoned with nigella, which lends its characteristic nutty flavor, and then garnished with fresh cilantro and mint.

THE COOK'S NOTES

- Because this recipe uses different types of vegetables, their average cooking times vary and there is a risk of some drying out before others are done. Rather than create both a confusing schedule for when to remove or add which vegetable to the baking sheet and a scary dishwashing situation, the vegetables are cooked covered in foil. This helps them cook in their own steam, and they won't dry out.

- Skip watery vegetables like tomatoes; those are better off being broiled or skewered on kebabs.

336

Master Mushroom Vegetable Stock

MAKES 3½ TO 4 CUPS [830 TO 945 ML]

Crush and place **3 oz [85 g] dried shiitake mushrooms** and **6 cups [1.4 L] water at 158°F [70°C]** (see the Cook's Notes) in a large heatproof mixing bowl or measuring cup and let sit for 30 minutes. Squeeze the mushrooms to release as much liquid as possible and discard the mushrooms. This is the mushroom "tea."

Heat a large Dutch oven, stockpot, or large, deep saucepan over medium heat. Do not add water or oil; we're going to use the liquid from the vegetables to braise and brown them. Add to the pot **1 large onion or 4 shallots (about 10½ oz [300 g]), thinly sliced**, and **1 large leek, white part only (about 9¾ oz [275 g]), thinly sliced**. Cover and cook, stirring occasionally to prevent burning, until the vegetables turn light brown, 6 to 8 minutes.

Add **1 lb [455 g] carrots, diced**, and **2 large celery stalks (about 6 oz [170 g]), thinly sliced**.

Cover and cook dry, stirring often and lowering the heat as necessary to avoid scorching, until the vegetables are tender and lightly browned, 30 to 45 minutes. If the onions start to burn, add 1 to 2 Tbsp water.

Stir in the mushroom tea, scraping the bottom of the pot to loosen any sticking vegetables. Add **1 Tbsp white or yellow miso paste**. Increase the heat to high and bring to a boil. Remove from the heat and let sit for 30 minutes.

Line a fine mesh sieve with a layer of cheesecloth and set it over a large bowl. Strain the liquid through the cheesecloth and gently squeeze the solids in the cloth to extract as much liquid as possible. Use the stock as needed or divide and store in the freezer for up to 3 months.

This recipe is adapted from *The Flavor Equation* cookbook and is an essential one to file away. I use it often to make soups, flavor rice for pulaos, and even steep vegetables. You can also drink a bowl on its own.

THE COOK'S NOTES

- The temperature is listed in the instructions here because at 158°F [70°C], the enzymes responsible for glutamate production in dried shiitake work most efficiently.

- I don't add salt to the stock until I use it.

- Avoid adding dried onion skins, as they'll make the stock taste bitter because they contain concentrated amounts of quercetin, a bitter-tasting polyphenol.

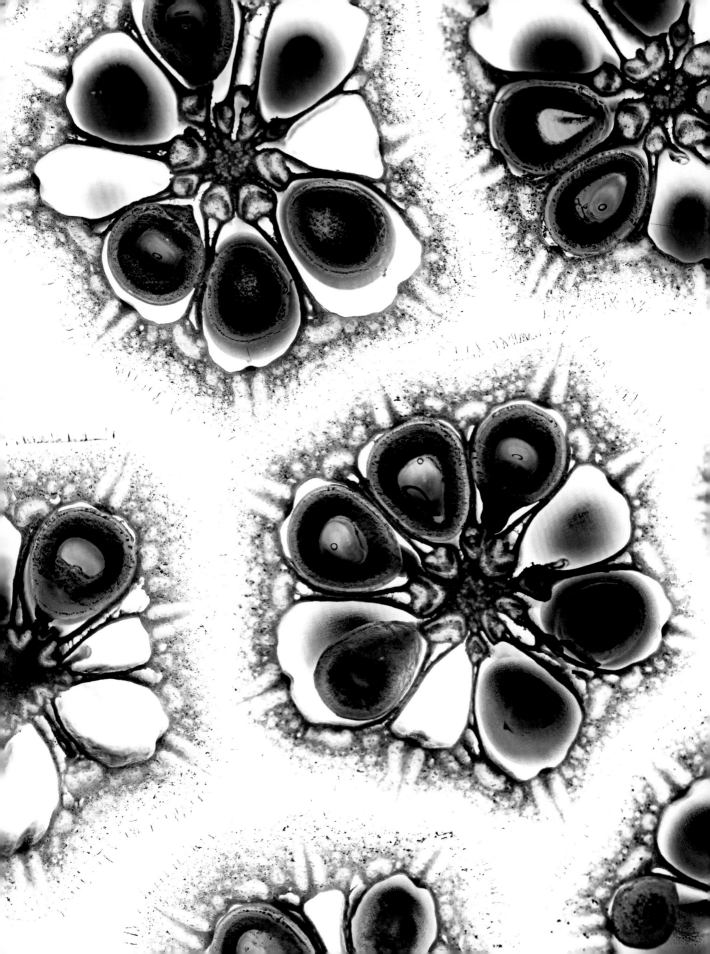

Platters, Boards + Tricks

Practically any dip or sauce from the recipes in this book can be repurposed to accompany a vegetable platter or board. This is one place where you should go wild and experiment with lots of different ideas.

Tips for Preparing a Platter or Board

- Select a theme for your platters and boards. It could be Indian, Middle Eastern, rainbow colored, *Star Wars*, and so on. The options are endless.

- Select the freshest vegetables: They should be crisp, bright, and full of life, and not have one leg in the afterlife.

- Go bold! Do not shy away from color, so grab those purple carrots and lay out those gorgeous Easter Egg radishes and Chioggia beets. Your farmers' market is your best friend; the vendors there grow and sell some of the most special vegetables.

- Make sure vegetables of the same kind are cut to the same dimensions or it will look a little haphazard.

- It's best to cut vegetables the day of for their best texture. To prevent cut vegetables from drying out, wrap the platter with a tight layer of cling film.

- Dips can be kept fresh by storing them covered tightly with a lid or a layer of cling film. For dips that contain olive oil, like the Peanut Muhammara (page 275), you can stir in 1 or 2 Tbsp olive oil and drizzle a bit extra on top to give it a refresh.

- In addition to small bowls of flaky salt and coarsely ground black pepper, I like to serve small bowls of spice blends like dukka, if they're a part of the overall theme of the platter.

- If serving cheeses, make sure they pair well with the vegetables and dips in your platter. Feta, grilled kefalotyri, and chanakh are some of my favorites to use with Mediterranean- and Middle Eastern–themed platters.

- You will need some tools to make it easy for your guests, such as wooden picks, labels for less familiar cheeses and dips, small tongs, cheese knives and forks, colorful napkins, and serving plates.

A Few Dips, Sauces, and Spreads to Choose From

- Avocado Caesar Dressing (page 120)
- Butter "Chicken" Sauce (page 255)
- Buttermilk Caraway Sauce (page 36)
- Cashew Green Chutney (page 93)
- Cashew Red Pepper Dip (page 123)
- Peanut Muhammara (page 275)
- Pistachio Pesto (page 310)
- Pumpkin Seed Chutney (page 190)
- Pumpkin Seed Sauce (page 258)
- Roasted Bell Pepper Sauce (page 299)
- Sauce Gribiche (page 87)
- Sweet Fennel Butter (page 70)
- Tomato Chutney (page 253)

A Few Recipes That Will Work Great on a Platter or Board

- Bombay Potato Croquettes (page 266)
- Saffron Lemon Confit with Alliums + Tomatoes (page 38)
- Steamed Artichokes with Cashew Red Pepper Dip (page 123)

I like to serve boards with small bowls filled with spiced fig jam or a hot and sweet pepper jelly, slices of Manchego or Gouda, spiced nuts, and slices of a baguette or crackers.

340

Spice Blends

You can buy spice blends from a store, but for those of you who want to make your own, here are my versions. Mix and bottle up!

Chaat Masala

MAKES ABOUT ¼ CUP [25 G]

This is the star spice blend of chaat, a beloved category of Indian street food. When ready to use, add ¼ tsp ground kala namak (Indian black salt) for every 1 tsp chaat masala.

Heat a small, dry cast-iron or stainless-steel skillet over medium-high heat. Turn down the heat to medium-low and add **2 tsp carom or ajwain seeds, 2 tsp cumin seeds, 2 tsp coriander seeds, 4 dried whole Kashmiri chillies** (see the Cook's Notes), **4 whole cloves, 1 tsp amchur (dried ground unripe mango powder), 1 tsp whole black peppercorns,** and a tiny **pinch of asafetida**. Toast by rotating the skillet to swirl the spices, until they turn fragrant and the cumin and coriander start to turn light brown. Remove from the heat and transfer to a plate to cool to room temperature. If the spices burn, discard and start fresh.

Add the cooled, toasted spices to a spice grinder or coffee mill along with **1 tsp dried mint** and **1 tsp ground dried ginger**. Blend on high speed until smooth. Store the spice blend in an airtight container at room temperature for up to 2 months.

Garam Masala

MAKES ABOUT ¼ CUP [25 G]

An essential spice blend in Indian and many South Asian cuisines, garam masala is used to flavor both meat and vegetable dishes.

Heat a small, dry cast-iron or stainless-steel skillet over medium-high heat. Turn down the heat to medium-low and add **2 Tbsp whole cumin seeds, 2 Tbsp whole coriander seeds, 1 Tbsp whole black peppercorns, one 2 in [5 cm] cinnamon stick, 12 whole cloves, 1 whole black cardamom pod, 3 or 4 whole green cardamom pods,** and **1 tsp freshly grated nutmeg**. Toast by rotating the skillet to swirl the spices, until they turn fragrant and the cumin and coriander start to turn light brown. Remove from the heat and transfer to a plate to cool to room temperature. If the spices burn, discard and start fresh.

Add the cooled, toasted spices to a spice grinder or coffee mill and blend on high speed until smooth. Store the spice blend in an airtight container at room temperature for up to 6 months.

Za'atar

MAKES ¼ CUP [50 G]

This is one of my favorite Middle Eastern spice mixes. Here's my version. I add salt when I'm about to use it.

In a small bowl, combine **2 Tbsp toasted white sesame seeds; 1 Tbsp dried oregano; 1 Tbsp dried thyme; 1 Tbsp ground cumin; 1 Tbsp red pepper flakes such as Aleppo, Maras, or Urfa;** and **1 tsp ground black pepper**. Transfer to an airtight container and store at room temperature for up to 4 months.

THE COOK'S NOTES
You can use either whole Kashmiri chillies or 1 Tbsp ground Kashmiri chillies. Another option is 1 tsp Kashmiri chilli powder substitute = ¾ tsp smoked sweet red paprika + ¼ tsp ground cayenne.

Store spices and spice blends away from light in a cool, dark spot in your pantry.

With Gratitude

A cookbook is more than just a book. The process is long and takes a few years to reach fruition. Recipes, photography, illustration, and design require input and the support of many people who work tirelessly behind the scenes to bring a book to life.

This book that sits in your hands exists due to the combined efforts of several special people.

To Sarah Billingsley, my editor, and Lizzie Vaughan, my book designer at Chronicle Books, and their entire team who've been instrumental in helping me push the boundaries with this project. To Matteo Riva, whose gift it was to translate my messy sketches into gorgeous artwork and make science look cool! Thank you for giving the geeky cook a place to write.

To the inimitable Maria Ribas, my literary agent, and the entire team at the Stonesong agency for being my champions at every stage of the book process. Maria, without you, none of my books would exist.

To my recipe tester, Lisa Nicklin, who tirelessly worked with me through all the recipes. Thank you for sharing your expertise and knowledge; it has made me a more conscious, wiser cook and writer.

To Daniel Gritzer, Sho Spaeth, and Emma Laperruque, who've been immensely helpful with any questions and were my sounding board on all topics related to vegetables. To my friends and fellow writers who've shared their knowledge with me and some of whom sat through numerous taste tests: Reem Kassis, Jacki Glick, and Akshay Mehta.

To the readers of my blog, *A Brown Table/Nik Sharma Cooks*; my newsletter, *The Flavor Files*; and my columns. Thank you for supporting my work; it is a privilege and a joy to see you cook my recipes and share them. Your feedback over the years helped shape the idea and the recipe style of this book.

Last but not least, my husband, Michael, who has to taste everything and knows the only thing he can complain about is the texture or taste, but not the number of times he has to eat the same dish over and over again during the phase of recipe testing.

—NIK

Sources Consulted + Recommended Reading

Vegetable Books

Vegetable Production and Practices by Gregory E. Welbaum (CABI, 2015)

Food Chemistry, 3rd edition, by H.-D. Belitz, W. Grosch, and P. Schieberle (Springer, 2009)

Plant Evolution under Domestication by Gideon Ladizinsky (Kluwer Academic Publishers, 1998)

Cookbooks

The Professional Chef by the Culinary Institute of America (Wiley, 2011)

The Flavor Equation by Nik Sharma (Chronicle Books, 2020)

The Food of Sichuan by Fuchsia Dunlop (Norton, 2019)

Cool Beans by Joe Yonan (Ten Speed Press, 2020)

The Classic Italian Cookbook by Marcella Hazan (Knopf, 1976)

Jane Grigson's Vegetable Book (Bison Books, 2007)

Japan: The Cookbook by Nancy Singleton Hachisu (Phaidon Press, 2018)

The Complete Vegetarian Cookbook by America's Test Kitchen (America's Test Kitchen, 2015)

Roots: The Definitive Compendium by Diane Morgan (Chronicle Books, 2012)

Vegetable Kingdom: The Abundant World of Vegan Recipes by Bryant Terry (Ten Speed Press, 2020)

Vegetable Literacy by Deborah Madison (Ten Speed Press, 2013)

Wine Folly: The Essential Guide to Wine by Madeline Puckette and Justin Hammack (Avery, 2015)

Websites

The Flavor Files
https://niksharma.substack.com

USDA/U.S. Department of Agriculture Food Data Central
https://fdc.nal.usda.gov/fdc-app.html

Post Harvest Center, University of California, Davis
https://postharvest.ucdavis.edu/Commodity_Resources/

The Kitchen Scientist at Food 52
https://food52.com/tags/the-kitchen-scientist

Asparagus

"We Unravel the Science Mysteries of Asparagus Pee," Angus Chen, *Food for Thought*, NPR, December 14, 2016, https://www.npr.org/sections/thesalt/2016/12/14/505420193/we-unravel-the-science-mysteries-of-asparagus-pee.

"Genetics of Asparagus Smell in Urine," Afsaneh Khetrapal, BSc, *News Medical Life Sciences*, March 2, 2021, https://www.news-medical.net/health/Genetics-of-Asparagus-Smell-in-Urine.aspx.

Beets

"Beeturia: The Myth," *Myths of Human Genetics*, John McDonald, University of Delaware (n.d.), https://udel.edu/~mcdonald/mythbeeturia.html.

Artichokes

"Pairing Flavours and the Temporal Order of Tasting," Charles Spence, Qian Janice Wang, and Jozef Youssef, *Flavour* (2017) 6:4, DOI 10.1186/s13411-017-0053, https://flavourjournal.biomedcentral.com/track/pdf/10.1186/s13411-017-0053-0.pdf.

Sunchokes

"Heat-Induced Degradation of Inulin," A. Böhm, I. Kaiser, A. Trebstein, and T. Henle, *Proc. Chemical Reaction in Food V*, Prague, September 29 to October 1, 2004, https://www.agriculturejournals.cz/publicFiles/236264.pdf.

Yams

"Yam" in World Vegetables, M. Yamaguchi, (Springer, 1983). https://doi.org/10.1007/978-94-011-7907-2_12.

"Yam Genomics Supports West Africa as a Major Cradle of Crop Domestication." N. Scarcelli, P. Cubry, R. Akakpo et al., Science Advances 5 (2019), DOI: 10.1126/sciadv.aaw1947.

"The Difference between Yams and Sweet Potatoes Is Structural Racism," Margaret Eby, Food and Wine, October 14, 2022. https://www.foodandwine.com/vegetables/the-difference-between-yams-and-sweet-potatoes-is-structural-racism.

Beans

"The Effect of Slow-Cooking on the Trypsin Inhibitor and Hemagglutinating Activities and in vitro Digestibility of Brown Beans (*Phaseolus vulgaris*, var.*Stella*) and kidney beans (*Phaseolus vulgaris*, var.*Montcalm*)," Monika Lowgren and Irvin E. Liene, *Plant Foods for Human Nutrition* 36 (1986): 147–154, https://link.springer.com/article/10.1007/BF01092141.

"The Degradation of Lectins, Phaseolin and Trypsin Inhibitors during Germination of White Kidney Beans, Phaseolus vulgaris L," F. H. Savelkoul, S. Tamminga, P. P. Leenaars, J. Schering, D. W. Ter Maat, Plant Foods for Human Nutrition 45(1994): 213–22. DOI: 10.1007/BF01094091.

Cauliflower

"Flower Development: Origin of the Cauliflower," David R. Smyth, *Current Biology* 5 (4) (April 1995): 361–363, https://www.sciencedirect.com/science/article/pii/S0960982295000728.

Serious Eats

Beans and Aquafaba Meringue

"The Science Behind Vegan Meringues," Nik Sharma, *Serious Eats*, May 19, 2021, https://www.seriouseats.com/science-of-aquafaba-meringues-5185233.

Kenji's Roasted Potatoes

"The Best Crispy Roast Potatoes Ever Recipe," J. Kenji López-Alt, *Serious Eats*, March 07, 2022, https://www.seriouseats.com/the-best-roast-potatoes-ever-recipe.

343

Index

A

347

W

Y

Z

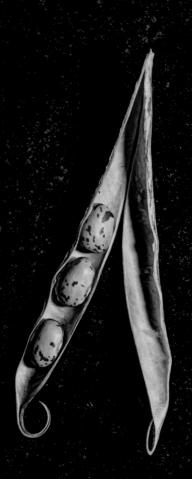

Photo by Michael Frazier

NIK SHARMA is the writer, photographer, and recipe developer behind *A Brown Table*, an award-winning blog that has garnered best-ofs from *Saveur*, *Parade*, *Better Homes & Gardens*, and the International Association of Culinary Professionals. Sharma lives in Los Angeles, California.

ALSO AVAILABLE: *The Flavor Equation* and *Season*

The Flavor Equation was named one of the best cookbooks of the year by the *New York Times*, *Eater*, *Epicurious*, *Food & Wine*, *Forbes*, *Saveur*, *Serious Eats*, the *Smithsonian* magazine, the *San Francisco Chronicle*, the *Los Angeles Times*, the *Boston Globe*, the *Chicago Tribune*, *CNN Travel*, *The Kitchn*, *Chowhound*, NPR, The Art of Eating Longlist 2021, and many more. It received international media attention from the *Financial Times*, the *Globe and Mail*, the *Telegraph*, the *Guardian*, the *Independent*, the *Times* (UK), *Delicious* magazine (UK), the *Irish Times*, and *Vogue India* and was the winner of the Guild of UK Food Writers (General Cookbook). It was a finalist for the 2021 IACP Cookbook Award.

Season was named one of the best cookbooks of the year by the *New York Times*, *NPR*, the *Guardian*, and *Bon Appétit*, was selected as an Amazon Book of the Month, and was a finalist for the 2018 James Beard Award.

Chronicle Books publishes distinctive books and gifts. From award-winning children's titles, bestselling cookbooks, and eclectic pop culture to acclaimed works of art and design, stationery, and journals, we craft publishing that's instantly recognizable for its spirit and creativity. Enjoy our publishing and become part of our community at www.chroniclebooks.com.

THE ASPARAGUS FAMILY
Asparagaceae

A
Asparagus

page 83

THE GRASS FAMILY *Poaceae*

Ba
Bamboo

page 63

Co
Corn

page 63

THE YAM FAMILY
Dioscoreaceae

Ya
Yams
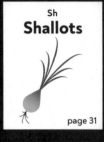
page 53

THE AMARYLLIS FAMILY *Amaryllidaceae*

Ch
Chives
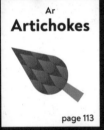
page 31

Ga
Garlic

page 31

Le
Leeks

page 31

On
Onions

page 31

Sc
Scallions

page 31

Sh
Shallots
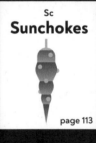
page 31

THE SUNFLOWER FAMILY *Asteraceae*

Ar
Artichokes

page 113

En
Endive

page 113

Es
Escarole

page 113

Lc
Lettuce

page 113

Rd
Radicchio

page 113

Sc
Sunchokes

page 113

THE MALLOW FAMILY
Malvaceae

O
Okra

page 249

THE GOURD FAMILY *Cucurbitaceae*

Cy
Chayote

page 185

Cu
Cucumber

page 185

Pu
Pumpkin

page 185

Sq
Squash

page 185

THE PEA OR BEAN FAMILY *Fabaceae or Leguminosae*

B
Beans
page 219

Cp
Chickpeas
page 219

Ji
Jícama

page 219

Le
Lentils

page 219

P
Peas
page 219

Sb
Soybeans
page 219